GOVERNING PRACTICES

Neoliberalism, Governmentality, and the Ethnographic Imaginary

Edited by Michelle Brady and Randy K. Lippert

Neoliberalism is one of the key analytical concepts that social scientists use to make sense of contemporary social change, with a third of all papers in anthropology and sociology using the term to label, explain, and critique transformations in political and everyday life. As an analytic category neoliberalism is powerful because it illuminates interconnections among diverse social and political changes occurring at multiple scales, from global financial regulations to everyday interactions with bureaucracies. Yet there are recurrent concerns that labelling political and social change "neoliberal" or "post-neoliberal" obfuscates more than it enlightens and encourages monolithic or totalizing assessments of change.

This volume moves forward these important debates about whether neoliberalism is a useful or appropriate analytic framework and highlights the best ways to deploy the concept. It presents a series of studies that serve as models for approaching an understanding of concrete practices of governance within specific domains. The contributors – a who's who of experts on neoliberalism – are united by the common influence of Foucault's governmentality approach and an ethnographic imaginary. Simultaneously the volume showcases key trends in contemporary governance, including governance through community; urban governance; rationalities of risk, resilience, and reconciliation; and governance through philanthropy.

MICHELLE BRADY is a research fellow in the School of Social Science at the University of Queensland.

RANDY K. LIPPERT is a professor in the Department of Sociology, Anthropology and Criminology at the University of Windsor.

Governing Practices

Neoliberalism, Governmentality, and the Ethnographic Imaginary

Edited by Michelle Brady and Randy K. Lippert

UNIVERSITY OF TORONTO PRESS
Toronto Buffalo London

© University of Toronto Press 2016
Toronto Buffalo London
www.utppublishing.com
Printed in the U.S.A.

ISBN 978-1-4875-0083-2 (cloth)   ISBN 978-1-4875-2061-8 (paper)

♾ Printed on acid-free, 100% post-consumer recycled paper with vegetable-based inks.

Library and Archives Canada Cataloguing in Publication

Governing practices : neoliberalism, governmentality, and the ethnographic imaginary/edited by Michelle Brady and Randy K. Lippert.

Includes bibliographical references.
ISBN 978-1-4875-0083-2 (cloth).   ISBN 978-1-4875-2061-8 (paper)

1. Neoliberalism.   I. Lippert, Randy K., 1966–, author, editor
II. Brady, Michelle, 1976–, author, editor

HB95.G69   2016   330.12'2   C2016-903872-6

This book has been published with the help of a grant from the Federation for the Humanities and Social Sciences, through the Awards to Scholarly Publications Program, using funds provided by the Social Sciences and Humanities Research Council of Canada.

University of Toronto Press acknowledges the financial assistance to its publishing program of the Canada Council for the Arts and the Ontario Arts Council, an agency of the Government of Ontario.

Canada Council **Conseil des Arts**
for the Arts **du Canada**

ONTARIO ARTS COUNCIL
CONSEIL DES ARTS DE L'ONTARIO
an Ontario government agency ·
un organisme du gouvernement de l'Ontario

Funded by the   Financé par le
Government   gouvernement
of Canada   du Canada

Contents

Acknowledgments

The editors would like to give sincere thanks to our editor Douglas Hildebrand for all his assistance with bringing this collection to publication. They wish to thank Laryssa Brooks for all her reliable professional work proofreading the volume. They thank *Foucault Studies* for permission to republish some material from a 2014 special issue, "Ethnographies of Neoliberal Governmentalities" (Vol. 18). In particular, Tania Murray Li's chapter in this volume is republished in full from this issue. Further, some material from the chapters by Randy Lippert, Michelle Brady, and by Katharyne Mitchell and Chris Lizotte also first appeared in that issue.

Michelle Brady acknowledges the Social Sciences and Humanities Research Council, and the following centres and departments at the University of Victoria: the Centre for Co-operative and Community-Based Economy, the Faculty of Human and Social Development, and the Department of Political Science, for providing funding in support of the November 2012 workshop where a number of the contributions to this volume were first presented. She would like to thank Julia Diamond, who provided outstanding assistance with organizing this workshop, Hiroyuki Ota and Aaron Duff for research assistance for this volume, and Louise Brady for valuable advice on formatting. Finally, she would like to warmly thank Cosmo Howard, Andrew Clarke, and members of the University of Queensland social theory reading group for conversations that were important in shaping the direction of the volume.

Randy K. Lippert acknowledges, as always, the loving support and patience of his partner, Francine, while co-editing this volume. Heartfelt thanks to Anna Pratt, James W. Williams, Kevin Walby, and Jesse

Seary for being great friends/colleagues from afar and for their special ways of helping bring this volume to fruition, sometimes just by listening. Special thanks too to Thomas Bud, perhaps the most efficient research assistant ever encountered, for reading over chapters at a crucial juncture; to Rhys Steckle for his diligence and incredible research skills for the chapter on condo life.

Contributors

Akin Akinwumi is a Postdoctoral Fellow in the Department of Geography at Simon Fraser University, Canada. His research interests have a broad interdisciplinary orientation, connecting human geography imaginaries with political studies, socio-legal studies, and social theory. These interests provide a route for empirical research on the rationalities and politics of reconciliation and nation-building in settler societies such as South Africa, Canada, and Australia. He is the author of "Powers of Reach: Legal Mobilization in a Post-Apartheid Redress Campaign" (*Social & Legal Studies* 2013) and "The Will to Transform: Nation-Building and the Strategic State in South Africa" (*Space and Polity* 2013).

Nicholas Blomley is a Professor of Geography at Simon Fraser University, Canada, specializing in legal geography, with a particular interest in the spatiality of property. He is the author of numerous articles. Most recently, he co-edited *Expanding Spaces of Law: A Timely Legal Geography* (Stanford, 2014). He is also the author of *Rights of Passage: Sidewalks and the Regulation of Urban Flow* (Routledge, 2011) and *Unsettling the City: Urban Land and the Politics of Property* (Routledge, 2004).

Michelle Brady is a Research Fellow in Sociology in the School of Social Science at the University of Queensland, Australia. Her research interests are in the area of interpretive policy studies and critical family studies. She has recently published in *Critical Policy Studies*; the *International Journal of Social Research Methodology*; *Work, Employment & Society*; the *Journal of Family Issues*; and *Foucault Studies*. She is an Associate Editor of the *Journal of Family Studies* and has been a guest editor for special issues of the *Journal of Family Studies* (2015) and *Foucault Studies* (2014).

Sara Dorow is an Associate Professor of Sociology at the University of Alberta, Canada, with teaching and research interests in mobility and migration; race, gender, and family; qualitative methodologies; and the politics of community, especially in resource production zones. She is the author of *Transnational Adoption: A Cultural Economy of Race, Gender and Kinship* (NYU Press, 2006), co-editor of a special issue of the *Canadian Journal of Sociology* (2013), and author or co-author of twenty articles, book chapters, and research reports. In 2009 she received the Distinguished Early Academic Career Award from the Confederation of Alberta Faculty Associations.

Cosmo Howard is a Senior Lecturer in the School of Government and International Relations and a member of the Centre for Governance and Public Policy at Griffith University, Australia, specializing in comparative public management, autonomous state agencies, and theories of individualization. He has published widely on administrative reform, policy making, and frontline public services. He is the co-author of *Social Policy, Public Policy: From Problem to Practice* (Allen & Unwin, 2001) and *The Service State: Rhetoric, Reality and Promise* (University of Ottawa Press, 2010). He is the editor of *Contested Individualization: Debates about Contemporary Personhood* (Palgrave Macmillan, 2007).

Wendy Larner is the Provost at Victoria University of Wellington, New Zealand. Her research is situated in the fields of globalization, governance, neoliberalism, and gender. Recent publications include a monograph, *Fashioning Globalization: New Zealand Design, Working Women and the Cultural Economy* (with Maureen Molloy, 2013), "Life, Ecology, Bodies: New Materialism and Feminisms" (*Feminist Theory* special issue, 2014), and "New Times, New Spaces: Gendered Transformations of Economy, Governance and Citizenship" (*Social Politics*, special issue, 2013), with Maria Fannin, Julie MacLeavy, and Winnie Wang. Wendy serves on the advisory boards of eight journals, including *Progress in Human Geography, Economy and Society, Environment and Planning A*, and *Studies in Political Economy*, and is a trustee of the Antipode Foundation. She is an Honorary Fellow of the Royal Society of New Zealand, a Fellow of the New Zealand Geographical Society, and a Fellow of the UK Academy of Social Sciences.

Tania Murray Li is a Professor of Anthropology at the University of Toronto, Canada, where she holds a Canada Research Chair in the

Political Economy and Culture of Asia. Her publications include *Land's End: Capitalist Relations on an Indigenous Frontier* (Duke University Press, 2014), *Powers of Exclusion: Land Dilemmas in Southeast Asia*, with Derek Hall and Philip Hirsch (NUS Press, 2011), the *Will to Improve: Governmentality, Development, and the Practice of Politics* (Duke University Press, 2007) and many articles on land, development, resource struggles, community, class, and indigeneity, with a particular focus on Indonesia.

Randy K. Lippert is a Professor of Sociology and Criminology at the University of Windsor, Canada, specializing in security, policing, and urban governance. He is the author or co-author of more than seventy refereed articles and book chapters and is the co-editor of *National Security, Surveillance and Terror: Canada and Australia in Comparative Perspective* (Palgrave Macmillan, 2016), *Corporate Security in the 21st Century* (Palgrave Macmillan, 2014), *Sanctuary Practices in International Perspective* (Routledge, 2014), *Policing Cities: Urban Securitization and Regulation in a 21st Century World* (Routledge, 2013), and *Eyes Everywhere: The Global Growth of Camera Surveillance* (Routledge, 2012). He is also the author of *Sanctuary, Sovereignty, Sacrifice: Canadian Sanctuary Incidents, Power, and Law* (UBC Press, 2006) and the co-author of *Municipal Corporate Security in International Context* (Routledge, 2015). He was Thinker in Residence at Deakin University, Australia, in 2015.

Chris Lizotte is a PhD candidate in the Department of Geography at the University of Washington, United States. His work centres on the evolution of neoliberal governance as it is manifested in contemporary school reform in the United States and France. In addition to his work at the University of Washington, Mr Lizotte has collaborated with research teams at the Institut français de géopolitique at the University of Paris 8 and the Géographie-Cités laboratory at the University of Paris 1.

Greg Marston is a Professor of Social Policy, School of Social Science at the University of Queensland, Australia. His main research interests include poverty, governance, and welfare state restructuring. He has authored more than sixty refereed articles and book chapters and is the International Editorial Adviser for the *Journal of Social Policy*. His recent co-authored books include the *Australian Welfare State: Who Benefits Now?* (Palgrave Australia, 2013) and *Work and the Welfare State: Street-Level Organizations and Workfare Politics* (Georgetown University Press, 2013). His co-edited book, *Basic Income in Australia and New Zealand:*

Perspectives from the Neoliberal Frontier (2016) was published with Palgrave Australia.

Katharyne Mitchell is a Professor of Geography at the University of Washington, United States, specializing in education, transnational migration, philanthropy, and urban governance. She has published over seventy-five articles and chapters and authored or edited nine books and guest-edited journals, including *Crossing the Neoliberal Line: Pacific Rim Migration and the Metropolis,* and *Practising Public Scholarship: Experiences and Possibilities beyond the Academy.* Mitchell served as Simpson Professor of the Public Humanities at the University of Washington from 2004 to 2007 and department chair from 2008 to 2013. She is the recipient of grants from the MacArthur Foundation, the Spencer Foundation, and the National Science Foundation.

Simon Moreton is an artist and researcher based at the University of the West of England, UK. His work considers the relationship between grassroots community and creative practice and broader political agendas. His PhD dissertation (University of Bristol, 2010) looked at affordable studio provision for artists in London, and he has gone on to research the Coexist project in Hamilton House (also at University of Bristol) and contemporary crafting communities (with the University of Exeter). He is currently a Research Fellow for the REACT hub, one of four initiatives established by the UK Arts and Humanities Research Council to fund knowledge exchange projects. His work has been published in *Environment and Planning A* and the *International Journal of Cultural Policy.*

Rob Shields is the University of Alberta's Henry Marshall Tory Endowed Research Chair in Sociology and a Professor of Sociology at the University of Alberta. His work spans architecture, planning, and urban geography. He is an award-winning author and a co-editor of numerous books, including *Spatial Questions* (Sage, 2013), *The Virtual* (Routledge, 2003), *Lifestyle Shopping* (Routledge, 1992), *Places on the Margin* (Routledge, 1991), and *Building Tomorrow: Innovation in Construction and Engineering,* co-edited with André Manseau (Ashgate, 2005), as well as online projects such as *Strip Appeal* and *Space and Culture.* Prior to moving to the University of Alberta he was Professor of Sociology and past director of the Institute of Interdisciplinary Studies at Carleton University, Canada. He was a Commonwealth Scholar at the University of Sussex, and his early career was in passive solar design, which he studied at Carleton University's School of Architecture. He founded *Space and Culture*, an international peer-refereed journal, and *Curb*, a Canadian planning magazine. He was 2014 City of Vienna Visiting Professor in Architecture and Planning at Technische Universität Wien and is currently completing research on nanotechnology as a space of concern.

Mariana Valverde is a Professor of Criminology at the University of Toronto, Canada. Her most recent book is *Chronotopes of Law: Scale, Jurisdiction and Governance* (Routledge, 2015). Previously, she published six books, six co-edited anthologies, and more than forty-five refereed scholarly articles. A key contribution to urban studies is her 2012 book *Everyday Law on the Street* (University of Chicago Press), which received the Law and Society Association's Herbert Jacobs Prize for best book of the year in sociolegal studies.

Openings

1 Neoliberalism, Governmental Assemblages, and the Ethnographic Imaginary

MICHELLE BRADY

Neoliberalism is among the most commonly used concepts in the social sciences.[1] As an analytic category neoliberalism is powerful in allowing scholars to illuminate interconnections among practices and changes occurring at multiple levels of governance, including links between the regulation of global finance and migration flows, everyday encounters in state and corporate bureaucracies, and individuals' self-understandings. Yet there are recurrent concerns that labelling social practices, policies, and shifts as "neoliberal" obfuscates more than it enlightens and encourages monolithic or totalizing assessments of our present and recent past (Flew, 2012; Rose, O'Malley, & Valverde, 2006). This volume seeks to contribute to and extend the current conversation about neoliberalism, especially as it relates to governmentality, by questioning common assumptions about its character as well as the extent to which it is shaping our lives.

One of the most influential and prolific groups of contributors to the literature on neoliberalism comprises those who use an analytics of governmentality framework, which is the focus of this volume. Foucault's small body of work on governmentality, which he produced in the late 1970s and early 1980s, was embraced by scholars in the English-speaking world keen to comprehend the politics of the "new right" in a more nuanced way. Foucault's reflections on governmentality attracted considerable attention during the 1990s with the publication of the anthology *The Foucault Effect: Studies in Governmentality* (Burchell, Gordon, & Miller, 1991), along with a series of articles instituting a new research program that built on Foucault's concept of governmentality (Miller & Rose, 1990; Rose & Miller, 1992).

In their seminal paper "Governing Economic Life," Miller & Rose (1990: 1) proposed a new approach to analysing governance and the

exercise of political power within advanced liberal democratic countries. They made the case for a focus on the ways language renders
"aspects of existence amenable to inscription and calculation" (2). A
particularly original aspect of their methodological approach lay in
their insistence that a focus on language as a technology should be
accompanied by an analysis of the "technologies of government" that
seek to grasp and shape the reality that rationalities have rendered as
an appropriate domain for governmental action (Barry, Osborne, &
Rose, 1996a: 11–5; Miller & Rose, 1990: 8). They argued (and many
have since agreed) that the strength of this approach lies in the way
that it highlights the diversity of mechanisms, groups, and forces
involved in governing beyond the state. Finally, their focus on the link
between governance and subjectification, particularly their claim that
within "advanced liberal democracies such as our own" technologies
of government "increasingly seek to act upon and instrumentalize the
self-regulating propensities of individuals in order to ally them with
sociopolitical objectives" (Miller & Rose, 1990: 28), has been enormously influential in terms of shaping dominant interpretations of
neoliberalism.

This volume endeavours to extend and refine the current conversation about neoliberalism by bringing together a diverse range of
scholars who use an analytics of governmentality to challenge and
extend existing research on neoliberalism. Studies of governmentality have traditionally relied exclusively on archival methods, but all
but one of the contributors to this volume use non-archival methods.
Regardless of whether they rely on archival or non-archival methods, all the contributors work with what I call, adapting Forsey's
(2010) term, an "ethnographic imaginary." Such an imaginary is
not a specific methodology or set of techniques, nor does it necessarily involve the traditional anthropological emphasis on participant observation. While some contributors draw on the kinds of
intense, long-term immersion in the field that the term "ethnography" implies, others do not draw on such sources. Such an imaginary
also does not involve adopting a realist epistemology[2] or relying on
traditional anthropological or sociological constructs such as society, state, or culture. Instead this imaginary involves reflecting upon
the particular geographic and temporal contexts within which practices or technologies of government unfold. It entails a sensitivity to
concrete practices within a milieu. Focusing on particular places and
times and using methods ranging from participant observation to

the analysis of transcripts of consultation hearings to archival methods, these scholars embrace an openness to the unexpected and seek to refine and trouble orthodox understandings of neoliberalism and broaden more widespread approaches to engaging in an analytics of governmentality.

A recurring theme in this collection is a focus on governmental assemblages (or ensembles) that link neoliberal with non-liberal rationalities and that involve heterogeneous practices and techniques, as we elaborate on in more detail later. For now it is sufficient to indicate that proponents of the concept of assemblage reject the idea noted above that merely because a program has neoliberal features it is "essentially neoliberal" (Rose et al., 2006: 98). For most it also involves rejecting the idea that neoliberal governmentality provides what Brenner, Peck, & Theodore (2010: 201–2) refer to as a "context of context – specifically, the evolving macrospatial frameworks and interspatial circulatory systems in which local regulatory projects unfold." However, this refusal to consider neoliberalism a "macrospatial" framework within which local projects unfold has been actively criticized by scholars adopting structural and political-economic approaches (Peck, 2013). An active debate has thus developed between, first, governmentality scholars who resist conceptual frameworks in which neoliberalism produces an overarching "context" (Collier, 2012); second, structural or political economy scholars (Peck & Theodore, 2012; Peck, 2013: 142) who imply that the "close-focus" of ethnographic and governmentality studies means that these are unable to address "issues related to the spatial patterning and historical evolution of neoliberal strategies and fronts *across* cases and contexts"; and third, scholars who argue that it is possible to combine governmentality and structural approaches by "giving oneself over to Foucault, in order to do Marx, better" (Fairbanks, 2012: 562). Contributors to this volume fall into the first and third camps, and thus the collection brings together a range of perspectives on this important debate.

This volume pushes imaginings of neoliberalism and ethnography into new spatial and temporal territories. It seeks to present fresh analyses of what is happening in myriad domains and to consider new ways of understanding these changes. Thus, it will be of value to the broad field of scholars interested in neoliberalism as well as to researchers engaged in an analytics of governmentality and the many scholars concerned with post-structuralist and ethnographic approaches to examining neoliberalism.

Neoliberalism

What is neoliberalism? Academic use of the term grew enormously during the 2000s, when it was linked to a stunning range of changes – everything from global financial deregulation to the rise of Bollywood films to the transformation of education (Boas & Gans-Morse, 2009; Hamann, 2009; Mudge, 2008). While many changes associated with neoliberalism are not new, the term itself is. During the 1990s researchers more commonly labelled the significant economic and/or political changes of the late 1970s and 1980s as "advanced liberalism" (Barry, Osborne, & Rose, 1996b; Rose, 1996, 1999), the rise of the "new right" (King, 1987; Levitas, 1986), or "economic rationalism" (Pusey, 1989). It was only during the 2000s that academics and activists came to commonly use "neoliberalism" (Boas and Gans-Morse, 2009: 138; Flew, 2014: 49–50; Kipnis, 2007: 383; Peck, Theodore, & Brenner, 2010: 97).[3]

Neoliberal processes of transformation are commonly identified as involving an emphasis on financial and labour market deregulation (Brenner & Theodore, 2002; Hilgers, 2010), a focus on the promotion of choice in the delivery of public services (Rose, 1999), "responsibilization" (placing more responsibility for this or that practice onto individuals) (Dean, 2010; Hilgers, 2010; Rose, 1999), and privatization or marketization of a range of government services (Brenner & Theodore, 2002; Dean, 2010).

However, there are also considerable areas of disagreement. While some scholars argue that neoliberalism is closely associated with efforts to radically limit state delivery of services (see Hilgers, 2010: 352 for a discussion of this view in anthropology) or intervention in everyday life, others suggest it is tethered to attempts to reconfigure and even expand state intervention (Ferguson, 2006; Harvey, 2007). Another area of disagreement concerns the relationship between neoliberalism and collectivism. Some suggest the former necessarily opposes the latter (Hilgers, 2010: 352). Other influential accounts of advanced liberalism suggest governmental programs commonly labelled as neoliberal actually promote certain kinds of collectivism, such as self-reliant communities (Rose, 1999: 135–6).

Addressing these kinds of questions is not simply a matter of acquiring more empirical data. Instead it requires being clearer about what kind of object neoliberalism is (i.e., its ontological status). Larner's (2000) seminal typology of neoliberalism as policy framework, ideology, and governmentality remains highly relevant to current reflections on this question. Yet more recent groupings developed by scholars such as Steger & Roy (2010), Hilgers (2010), Flew (2012), Collier (2012), and Wacquant (2012) bear consideration too. These scholars have argued that neoliberalism variously refers to a type of culture; a policy framework; a dominant ideology; a thing that determines everything else in the social field; a specific governmental rationality; or an assemblage of technologies, reflections, and so on.

Obviously, there is significant overlap in these ways of categorizing neoliberalism and not all approaches are equally predominant. Understandings of neoliberalism as cultural are among the earliest (e.g., Pusey, 1989), but also the most recent, additions to the literature (Hilgers, 2010). Neoliberalism as policy framework was one of the initial means of characterizing the radical economic, political, and policy upheavals of the late 1970s and 1980s, but it has become increasingly overshadowed by research that understands neoliberalism as a dominant ideology "disseminated by hegemonic economic and political groups" (Larner, 2000: 12; van Apeldoorn & De Graaff, 2012), and research that conceives of it as a form of governmentality or as a political rationality (Collier, 2012; Larner, 2000) (see below).

As Collier argues, those who view neoliberalism as an ideology regard it as "a macro-structure or explanatory background against which other things are understood," and they tend to portray "neoliberalism as bigger, stronger, more structural and structuring than other things in the field ... such that we can call the whole mess neoliberalism" (Collier, 2012: 186, 191). Scholars who characterize neoliberalism this way have sought to respond to criticisms that they depict neoliberalism as a totalizing configuration by developing new concepts such as "contingent urban neoliberalism" (Wilson, 2004), "variegated neoliberalization" (Brenner et al., 2010), and "roll-with-it neoliberalization" (Keil, 2011), which more explicitly recognize contingencies of neoliberal reforms (see Lippert in this volume). However, scholars such as Collier (2012: 189) have argued that "no matter how many modifiers" these scholars add to "neoliberal," it does not change the fact that the term still "designates phenomena at the level of structure, the context of context or the macro-context."

Many scholars embracing a governmentality approach reject structuralist approaches to neoliberalism, some on the grounds that they pay insufficient attention to the *logic* (or rationality) of non-neoliberal rationalities (see Lippert in this volume), while others seek to combine governmentality and more structuralist approaches (see Mitchell and Lizotte in this volume). The focus of this volume is on research that

8 Michelle Brady

understands neoliberalism as governmental rationality and it is this to which we now turn.

Governmentality

Studies of neoliberalism as governmentality or as a political rationality build upon Michel Foucault's work on governmentalities, political rationalities, and neoliberalism (1979, 1984, 1987, 1988, 2000a, 2008). Foucault argued that what distinguished all modern forms of political thought and action is a certain mentality for which he coined the term governmentality. Within his lecture on governmentality he defined it as three interrelated things. First, governmentality refers to "the ensemble formed by the institutions, procedures, analyses and reflections, the calculations and tactics that allow the exercise of this very specific albeit complex form of power, which has as its target population, as its principal form of knowledge political economy and as its essential technical means apparatuses of security" (1979: 20). Second, it involves the tendency over time for this type of power to dominate other forms of power, which resulted in the formation of a series of "state apparatuses" for government and "the development of a whole complex of 'savoir'" (ibid.). Finally, governmentality is the process through which the administrative state became governmentalized in the fifteenth and sixteenth centuries (ibid.).

Around the time of the publication of *Discipline and Punish* in the mid-1970s Foucault made clear his desire to avoid state-centred approaches to the study of power relations. However, following *Discipline and Punish*, Foucault desired to move beyond analysis of specific institutions to understand "the state's strategic role in the historical organization of power relationships"; his concept of governmentality enabled him to achieve this aim without falling back on a state-centred approach (Bröckling, Krasmann, & Lemke, 2011: 2; Donzelot, 1979: 77). During his lifetime Foucault never published a major work on governmentality, political rationalities, or neoliberalism (Donzelot & Gordon, 2009: 3), but the insightfulness of his small collection of writings on these topics has inspired a large body of work on liberal and neoliberal political rationalities (briefly outlined above). Contemporary scholars (and English-speaking scholars in particular) have overwhelmingly relied on a wide range of secondary accounts (cf. Bröckling et al., 2011: 9; Donzelot & Gordon, 2009) together with a handful of Foucault's own writings on governmentality.[4] Secondary accounts have included studies by Foucault's collaborators published in *The Foucault Effect* (Burchell et al., 1991); summaries and interpretations of Foucault's Collège de France lectures (Gordon, 1987, 1991; Lemke, 2001, 2002); and monographs on governmentality by sociologists such as Nikolas Rose and Mitchell Dean (Dean, 1994, 1999, 2010; Rose, 1999). For this reason we focus below primarily on secondary accounts and interpretations rather than on Foucault's own body of work.

Foucault's brief but perceptive reflections on contemporary governance and neoliberalism were quickly embraced by Anglo scholars (Donzelot & Gordon, 2009: 3) keen to make sense of the staggering political, policy, and economic changes that commenced in the late 1970s. These entailed the rise of the "new right" in the United Kingdom and North America and "economic rationalism" in Australia and New Zealand. Foucault's analytical emphasis on governance as the "conduct of conduct" and eschewal of the idea that state power emanates from a single source (the economy or the state) resonated with the emerging emphasis on politics "beyond the state," and new governance arrangements, including a seemingly new emphasis on public-private partnerships (see Valverde in this volume for an argument that these forms of partnerships were not, in fact, new). These new emphases emanated not only from the new right, but also from new feminist politics of personal life and new critiques of bureaucracy based on the idea of client rights. All these critiques gestured towards new ways of thinking about politics and governance (Yeatman, 1990) and challenged traditional divisions of freedom versus constraint, state expansion versus retraction, and public versus private. Scholars embraced Foucault's conceptualization of governance as governmentality because it enabled them to move beyond these staid divisions and grasp "new right" politics in more nuanced and detailed ways.

During the 1980s Thatcher's and Reagan's market-driven approaches to public policy overwhelmingly were read as a retraction of the state and/or a reversion to the classic liberalism of the nineteenth century. New studies of governmentalities, inspired by Foucault's analysis, argued they were better understood as "a restructuring of governmental techniques" (Bröckling et al., 2011: 9) and the emergence of a new political rationality that developed out of the practical critiques and failures of what Rose (1999) has called "social liberalism." Over the last twenty years an analytics of governmentality has fruitfully examined the transformation of governance in many fields and jurisdictions, including employment in Australia (Dean, 1998; Harris, 2001) and the United Kingdom (Walters, 1997), processes of privatization in New Zealand and the United Kingdom (Larner, 2000), neoliberalism and financial planning in Argentina and the United States (Binkley, 2009; Fridman, 2010), neoliberalism and development in Indonesia (Li, 2007a), the rise and development of school counselling in the United Kingdom and United States (Besley, 2002; Fejes, 2008), the reform of housing policy in the United Kingdom (Flint, 2003; McKee, 2009; Stenson & Watt, 1999), the governance of crime (O'Malley, 1992; Simon, 2007), security and surveillance (Haggerty & Ericson, 2000; Lippert, 2009), Canadian and international immigration and refugee regimes (Lippert, 1998, 1999; Lippert & Pyykkönen, 2012b), the relationship between neoliberalism and standardization (Higgins & Larner, 2010) and global geographies of neoliberalism (Larner, 2009; Larner & Walters, 2004) among many others.

Having established the growth and popularity of studies that are inspired by an analytics of governmentality, how was this analytic deployed and developed within these works? Within his highly influential book *Powers of Freedom* Rose argued: "To analyze ... through the analytics of governmentality is not to start from the apparently obvious historical or sociological question: what happened and why? It is to start by asking what authorities of various sorts wanted to happen, in relation to problems defined how, in pursuit of what objectives, through what strategies and techniques" (1999: 20).

In emphasizing how problems are defined, and the strategies and techniques used to achieve objectives, Rose foregrounds the concepts of problematization and technologies. These two concepts have been central to research on governmentality since Rose and Miller proposed a new analytics of governmentality that would provide insights into "political power beyond the state" through a focus on programs, rationalities, and technologies (1992). Consistent with Foucault's emphasis on conditions of possibility and on practices of governance, the analytics of governmentality described by Rose and Miller, and others elsewhere, focused on how problems had been defined (i.e., problematization) and also foregrounded the role that technologies played in governance.

The emphasis on problematization was articulated by Rose in the idiom of "political rationalities," which, he argues "are characterized by regularities" (1999: 26) and can be distinguished by identifying and contrasting their moral form (understanding of justice, freedom, etc.), their epistemological character (understanding of the objects they govern), and their idiom (the kinds of terms and phrases that are used) (26–7).

Governmentality studies also emphasized the way in which governmental techniques helped to produce certain patterns or issues as "problems" that governments or other actors could and should address (see also Lippert & Stenson, 2010) in addition to the more traditional emphasis on the techniques that actors use to address the problems they seek to govern. Rose's work, for example, first examined how technologies helped to produce governable spaces (e.g., by dividing time in a certain way), and governable subjects (by producing conceptions of subjectivity and technologies of the self) (1999: 45). Second, he emphasized that "thought becomes governmental to the extent ... it attaches itself to a technology for its realization" (52) and provided the example of the schoolroom, which was invented in the nineteenth century.

Within the early work on governmentality the role of technologies in governance was foregrounded, but as we elaborate in the following section, since Rose and Miller (1992) they have been generally assigned a less important place than rationalities (Lippert, 2010 also makes this argument). An analysis of technologies involves a concern with the ways government creates governable spaces (e.g., by dividing time buildings in a certain way), governable subjects (by producing conceptions of subjectivity and technologies of the self; see Rose, 1999: 45), and political technologies (such as the schoolroom). Although Rose argues that thought becomes governmental in so far as it connects to a technology that can realize it, he also states that technologies for governing are "never simply a realization of a programme, strategy or intention" (ibid.: 52). Rose used this analytical strategy to analyse major political, policy, and economic changes from the late 1970s onward, arguing that they shared many of the premises of the neoliberal thought of the 1940s and 1950s and "entailed new domains to which government must address itself – the market, the family, the community, the individual – and new ways of allocating the tasks of government between the political apparatus ... professionals, economic actors, communities and private citizens" (ibid.: 139–40). As many others have argued, this analytic strategy highlights practices of problematization and regimes of practices or technologies that come to be seen as self-evident solutions to such problems.

Recently, scholars have argued for greater attention to other practices and concepts, including the situated practice of thinking (Collier, 2009: 80) (as opposed to diagrams of power) and the practice of assemblage (Li, 2007b: 131, 263) together with a greater emphasis on the concept of assemblage more broadly (see below). We return to this point in the following section on the ethnographic imaginary, where we highlight how scholars undertaking ethnographic studies of governmentalities explicitly foreground a range of different practices of government.

Central to early work on governmentality, too, was the idea of being open to the unexpected. In part this was to be achieved through a concern with minor, obscure thinkers (including bureaucrats and administrators) and an interest in mundane and micro-governmental techniques and tools (such as diaries, brochures, and manuals) rather than an exclusive focus on the thought of famous philosophers and politicians (Rose et al., 2006: 89). Some have already documented that, as the governmentality literature grew, it developed tendencies that ran contrary to this openness and "experimental investigation" (Donzelot & Gordon, 2009; O'Malley & Valverde, 2014; Osborne in Bröckling et al., 2011: 16; Rose et al. 2006). As Rose et al. (2006: 97) have lamented, this powerful "capacity to render neoliberalism visible in new ways" has been accompanied by a less celebrated aspect, which is a "marked tendency within governmentality studies to regard... [neoliberalism] as a more or less constant master category that can be used both to understand and to explain all manner of political programs across a wide variety of settings." They further argue that an outcome of this inclination is a kind of "cookie-cutter typification or explanation," where any program with "neo-liberal elements" is identified as essentially neoliberal (97–8). Gordon has echoed this assessment of the governmentality literature on neoliberalism, arguing: "The reception of Foucault's analysis of neoliberalism unfortunately often seems to be flattened into a set of polemical, ideological and globalising generalities, dispensing with the kind of descriptive investigation Foucault undertook in 1979 of the different avatars of neoliberalism with their national, historical and theoretical specificities" (Donzelot & Gordon, 2009: 7).

Bröckling et al. (2011: 16) came to a similar conclusion, arguing that one variant of this tendency are empirical studies of policy or political transformation occurring within short time periods that "distil the always identical rationalities, strategies and technologies of neoliberalism," while another variant is studies that focus on longer time frames but rehearse familiar sketches of governmental transformations from "Polizei to liberalism, or from the welfare state to neoliberalism" (Osborne, cited in ibid.).

Researchers' methodological responses to these identified weaknesses have been diverse. Some have continued to rely on traditional archival and documentary methods, while seeking to avoid the kinds of errors identified by Donzelot & Gordon (2009), O'Malley & Valverde (2014), Bröckling et al. (2011), and Rose et al. (2006) through avoiding grand theory and instead developing their conclusions from careful analysis of the empirical material they are working with. Another response has been to embrace ethnographic or quasi-ethnographic methods to provide closer, more context-specific analysis, which is referred to in this volume as an *ethnographic imaginary*. Working across a range of disciplines, including anthropology, sociology, criminology, geography, and social work, these scholars focus variously on the politics of governance (Li, 2007a; Mitchell, 2006), non-liberal rationalities (Hoffman, 2010; Lippert, 2014), governmental assemblages (Li, 2007a, 2007b), thinking understood as a "situated practice of critical reflection" (Collier, 2009: 80), and the work required to assemble heterogeneous elements together (Li, 2007a, 2007b). These studies have developed very important new insights into contemporary governance and neoliberalism.

Ethnographic Imaginary

The idea that an analytics of governmentality can fruitfully be used together with more traditional social scientific methods, including interviews or observations, is not universally accepted. Nikolas Rose (1999: 19) has long argued that researchers must choose between studies of governmentalities, which are focused on programmatic thinking and techniques and technologies for the conduct of conduct, and sociologies of practices.[5] This has been taken to mean that those embracing an analytics of governmentality should focus on schemas and pronouncements using archival methods. Such a position largely aligns with Foucault's argument that his book *Discipline and Punish* did not aim to understand "living reality" or "real life," but instead aimed to "analyze the connection between ways of distinguishing between true and false and ways of governing oneself and others" (2000c: 233). Foucault stated that he acknowledged that "the actual functioning of the prisons ... was a witches' brew compared to the beautiful Benthamite machine" (232–3). Yet his analysis of plans for government or "theoreticians' schemas" was important, he argued, insofar as they were "fragments of reality that induce ... particular effects in the real" (233).

Not all scholars, however, have embraced this idea of focusing solely on "theoreticians' schemas" and using only archival methods. In the last decade researchers from across the social sciences have sought to put together an analytics of governmentality with methodologies (such as ethnography) or methods (such as interviews) more traditionally associated with realist social research (Caswell, Marston, & Larsen, 2010; Collier, 2011b; Lippert, 2014; Mitchell, 2006; Li, 2007a. Some scholars have done so without explicitly identifying for the reader how this research relates to traditional studies of governmentality that are based entirely on documentary evidence. Other scholars, such as Stephen Collier (2011b), Tania Murray Li (2007a) and Randy Lippert (2005) have shared with the reader their own methodological reflections on the contribution of ethnographic research to studies of governmentalities. Li has shared her view that the "study of the rationale of governmental schemes and the study of social history [are] distinct kinds of inquiry," while making her principal subject of analysis "the intersection of governmental programs with the world they would transform" (2007a: 27). Collier has shared a distinctly different approach in his book *Post-Soviet Social: Neoliberalism, Social Modernity*, where he argues that Foucault analysed "political reasoning as a situated practice through which existing governmental forms are reflected upon, reworked and deployed" (2011b: 19). Collier's project, which entailed fieldwork combined with extensive archival research, therefore sought to analyse all the actors that he encountered "as *thinkers*, and to analyze *thinking* ... as a practico-critical activity" (28; emphasis in the original). As Collier explains in a recent talk, doing so involves refusing to give "fieldwork in the form of ethnography ... any epistemic privilege" over other forms of knowledge, and rejecting a privileged role for it because "it gets close to practices, to experience, to the quotidian, the anecdotal, the local, the circumstantial" (2013: 13–14). At the same time he acknowledges that his fieldwork allowed him to observe "a grouping of sites and a set of problems" that would otherwise have remained invisible (29). Within his book on Canadian church sanctuary practices, Lippert takes a stance similar to Collier's, arguing first that rationalities are apparent in text as well as talk of sanctuary providers and second that adopting traditional sociological methods like interviews does not mean that "the object of inquiry automatically becomes 'the real'" (2005: 10). Like Collier, he argues that his use of fieldwork (including interviews and access to private archival material) revealed problems and discourses that were otherwise inaccessible. This issue will be returned to in the introduction to the chapters in this volume, below.

Assemblage/Ensembles/Apparatus

Assemblage is a reoccurring term in works that bring together an "ethnographic imaginary" and an analytics of governmentality. It is worth highlighting that, while the term assemblage is prominent in this volume, Foucault's term "dispositive" (apparatus) largely has been eschewed. Through "dispositive" Foucault sought to identify "a thoroughly heterogeneous ensemble consisting of discourses, institutions, architectural forms, regulatory decisions, laws, administrative measures, scientific statements, philosophical, moral and philanthropic propositions – in short, the said as much as the unsaid" (1980b: 194). He succinctly notes that "dispositive" referred to a "formation which has as its major function at a given historical moment that of responding to an *urgent need*" (195; emphasis in the original). As elaborated below, many scholars have suggested "dispositive" or apparatus signal closure and have turned instead to concept of assemblage, which is associated with Deleuze & Guattari (1987). As Lippert and Pyykkönen (2012a) have argued: "An 'assemblage' is a contingent and creative ensemble of distinctive material and social elements that can include knowledges, ways of seeing and calculating, human capacities, mundane and grand devices, kinds of authority, spatialities, and governmentalities ... that converge and which seek a specific outcome among those who govern and of those who are governed."

Other scholars argue that Foucault's later work is underappreciated in resolving his tendency to render relations of power as totalizing and fixed. For example, O'Malley & Valverde (2014: 323–5) contend that in later works Foucault pointed to how "forms of power [such as] 'government,' 'governmentality' and 'security'" formed "complex hybrids and articulations ... with discipline and sovereignty," and the ongoing "triangular" relationship between sovereignty, discipline, and government[ality] which produced "hybridizations, interactions, alliances, and so on." Along somewhat similar lines Collier (2009: 78–80) suggests that an important mutation occurred in Foucault's "method and diagnostic style" from 1976 to 1979, whereby he moved from "analysis ... couched in surprisingly epochal and totalizing" terms to a "topological analysis" focused on examining the "patterns of correlation in which heterogeneous elements – techniques ... institutional structures – are configured, as well as the redeployments through which these patterns are transformed."

However, many scholars, have found lacking Foucault's recognition of heterogeneity, contingency, and the decentred character of governance and have turned to the concept of assemblage for additional analytical resources. The term "assemblage" has been deployed in numerous fields and subfields, but our particular concern is how it has been taken up, elaborated on, and extended by those adopting the "ethnographic imaginary" and governmentality, namely, scholars such as Collier, Lippert, Li, Ong, and Rabinow. In the following brief review of how these scholars have used the concept of assemblage I aim to underscore that this notion has been adopted and developed in different ways.

Writing over a decade ago in *Anthropos Today*, Rabinow developed an analytical framework that distinguished between assemblages, problematizations, and apparatuses. He argues that the primary object of his recent inquiries has been "assemblages" that are "a distinctive type of experimental matrix of heterogeneous elements, techniques and concepts ... They are comparatively effervescent, disappearing in years or decades rather than centuries ... the temporality of assemblages is qualitatively different from that of either problematizations or apparatuses [which have a long duration]" (2003: 56).

Two years later, in their anthology *Global Assemblages*, Ong and Collier similarly used the concept (as Collier [2011a] explained in a later interview) to "emphasize a rather foreshortened time frame and very contingent relationships." They argued, "Global assemblages provide a framework for anthropological studies of globalization because "as a composite concept, the term "global assemblages" suggests inherent tensions: global implies broadly encompassing, seamless, and mobile; assemblage implies heterogeneous, contingent, unstable, partial, and situated" (Ong & Collier, 2005: 12).

Similarly within security and migration studies, Lippert and colleagues (Lippert & O'Connor, 2003; Lippert & Pyykkönen, 2012b) deployed an understanding of assemblage that drew from Foucault's concept of the social apparatus. An assemblage, they argue, is "an open-ended system of relations between diverse programmatic efforts ... [that] can be grafted from or grafted to various rationalities and technologies of rule [provided the elements work together and] ... can be assembled from diverse elements ... and can cut across private- and public-sectoral divisions" (Lippert & O'Connor, 2003: 333).

A different approach to assemblage has been taken by Tania Murray Li, whose work on governmentality, neoliberalism, and community forestry management has been very influential in the literature. As Legg argues, assemblage for Li is "an act of labour and governance" (2011: 131). Li's primary reason for embracing a focus on assemblage is that it allows her to "highlight some aspects of government thus far neglected" (2007b: 264), and the desire to focus on more contingent relationships is a secondary consideration. That is, although Li briefly notes that she embraces the term "assemblage" (drawn from the work of Deleuze) rather than "regime of practices" or "dispositif," because the latter terms focus attention on the resultant formation and imply a completed order (fn 3) her primary concern is to use "assemblage" to highlight particular practices of government that have been neglected. Studies of governmentalities, she argues, have focused too much on the practice of problematization, whereby particular issues come to be represented as problems and certain solutions to these problems are rendered as self-evident while others are marginalized (264). An "analytics of assemblage," in contrast, (1) draws attention to "the hard work required to draw heterogeneous elements together" and implies that sometimes they might not "be made to cohere"; (2) emphasizes the continuous work of pulling together of "disparate elements"; and (3) recognizes the work of "the situated subjects who do the work of pulling together disparate elements" (264–5).[6]

The concept of assemblage is particularly important to this volume because it provides a starting point for thinking about disparate elements involved in governing that include but are not limited to neoliberalism. Since assemblages are open-ended and heterogeneous, the assumption of this volume is that they at least lend themselves to ethnographic work broadly conceived. At the same time ethnographic work itself, we suggest, may require such a concept to supplement the more dominant concepts of problematization and the associated apparatus (or regimes). An "ethnographic imaginary" permits going further to see the end and the mottled collections of discourses and practices that may otherwise remain hidden and, if undertaken over a long period, perhaps also their comparative effervescence.

About This Volume

This edited volume presents a range of material and arguments, but it is closely accompanied by a cohesive set of themes and debates. Its central argument is that neoliberalism should not be viewed as totalizing or unchanging, but as having mutated over time, as varying across geographical spaces and as coexisting within broader assemblages with other logics (see Lippert in this volume) or with as yet unnamed rationalities (see Valverde in this volume). We think that, first, showing what is outside neoliberal governmentality is important, and second, that this needs to be thought of as more than vague resistance or obstruction. This volume traces the relationship between the ethnographic imaginary and the effort to better understand neoliberalism primarily through an analytics of governmentality perspective. Within some chapters an analytics of governmentality is combined with critical perspectives such as political economy (see Marston, and Mitchell and Lizotte in this volume) or remains more backgrounded or implied (see Valverde in this volume).

The collection presents works that question commonly accepted narratives about neoliberalism that embrace the notion of assemblage, that document how neoliberal rationalities are reshaping institutions and how we understand and act upon ourselves (subjectivities) by bringing together an analytics of governmentality with an ethnographic imaginary. However, these essays should not be viewed as examples of a single new methodology. Furthermore, the kind of ethnography they embrace does not necessarily involve participant observation or long-term fieldwork. Nor do the contributors depend on grand anthropological or sociological constructs such as society, state, or culture. Instead, their works involve diverse attempts to bring social science methods and concerns together with approaches informed by studies of governmentalities, to understand neoliberalism within a specific place and time, and to critically reflect on dominant narratives around neoliberalism. Currently, the works are scattered across disciplines including geography, sociology, political science, criminology, and anthropology, and they involve scholars working from very different traditions and problems. Our aim in bringing them together in an edited volume is to enable their accounts to reach a wider audience and to propel current debates forward by highlighting diversity and disagreement within this group of scholars and by toppling disciplinary barriers.

This volume does not attempt to reconcile or smooth over differences among perspectives. Instead we seek to merely highlight and identify points of difference for readers. The volume includes one anthropologist (Li), six geographers (Akinwumi, Blomley, Larner, Moreton, Mitchell, and Lizotte), two criminologists (Lippert and Valverde), three sociologists (Brady, Dorow, and Shields), and two policy/political scientists (Howard and Marston). Simultaneously, the volume showcases key trends in contemporary governance, including technologies and politics; governing through community; urban governance; rationalities of risk, resilience, and reconciliation; and new forms of philanthropy and welfare. It will be of value to the broad field of scholars interested in neoliberalism as well as to those scholars across the social sciences interested in governmentality studies and to the large number of anthropologists, sociologists, geographers, and other scholars interested in ethnographic approaches broadly conceived.

Most chapters were written for an international workshop held in November 2012 at the University of Victoria, British Columbia. Other prominent scholars, such as Nicholas Blomley, Rob Shields, and Mariana Valverde, were subsequently invited to contribute original chapters and graciously did so alone or with a colleague. Contributors thus include widely cited scholars in anthropology, criminology, geography, political science, and sociology,

Neoliberal Subjectivities? Examining Resilience, Investment, and Enterprise

One of the key areas that have been examined by studies embracing an "ethnographic imaginary" is what is understood as subjectivity. As Miller and Rose (1990: 18) argued, "governing operates through subjects." The idea that programs seek to act upon individuals' ability to steer themselves by shaping their desires and aspirations has been widely taken up and applied in the literature on contemporary neoliberalism. The subjectivities most consistently linked to neoliberal rationalities are *homo economicus* or, relatedly, the enterprising self that focuses all efforts on economic success (Rose, 1999). The chapters in this section engage with influential accounts of neoliberalism by showing that the subjectivities that are promoted within contemporary governance strategies, programs, and initiatives are remarkably varied, go beyond "economic man," and sometimes cannot be easily characterized as neoliberal.

In their chapter on resilience, the Coexist Project, and subjectivities, Wendy Larner and Simon Moreton challenge dominant accounts of resilience and subjectification. Using ethnographic methods informed by studies of governmentality, they examine a Bristol-based community interest company's attempts to foster resilient subjects and spaces. Named "Coexist" because of its aim to provide a physical space where people can work and live with one another and the broader environment, this project involves a wide range of persons, including artists, community groups, and health workers operating together to allow social organizations to flourish in a common space at a low cost. The participants in Coexist seek to construct broader networks of diverse groups and to serve as a model for others seeking urban revitalization and social innovation beyond Bristol. Larner and Moreton argue that Coexist seeks to produce collectivized subjects that are self-sustaining and empowered and are not the typical de-socialized, privatized, or prudential neoliberal subjects, who calculate the costs and benefits or risks of their actions.

Sara Dorow's chapter also takes up the issue of community and subjectivity. Focusing on community and on neoliberal governmentality in the oil sands zone of northern Alberta, Canada, Dorow considers Nikolas Rose's well-known formulation of community, which involves "reconstructing citizens as moral subjects of responsible community" (Rose, 2000: 1). Drawing from her ethnographic study of Fort McMurray, the key "urban service area" for the oil sands, Dorow argues that in Fort McMurray community becomes a moral project that all aspirational citizens are exhorted to strive for. Governing through community in the oil sands zone requires multiple government and non-government actors and three key strategies: conditioning investment in home, conditioning subjects to "give back" and conditioning investment in a community "beyond oil." Through the use of ethnographic methods Dorow identifies how actors beyond the state seek to govern through community, and she identifies the processes of inclusion and exclusion associated with such efforts.

Tania Murray Li's chapter also considers processes of exclusion and inclusion in relation to neoliberalism, governance, community, and subjectivity. Li investigates the role of expert knowledge in relation to governmentalities and highlights the ways that governmental assemblages in the global south since colonial times have sought to produce categories of cultural difference and to fix certain populations into non-market spaces. She argues that this practice actively continues today and is not simply a holdover from the past. However, compared with liberal assemblages, contemporary neoliberal assemblages operate with different understandings of the subjects that they seek to govern. While liberal colonial assemblages sought expert knowledge about the character of those who were governed, neoliberal assemblages assume *homo economicus* to be universal and thus seek expert knowledge regarding how to establish suitable incentives for the subjects whose conduct they seek to conduct. Within her chapter Li seeks to highlight the role ethnography has played in government and also to articulate the limits of an analysis that relies entirely on documentary/archival analysis.

Neoliberal Technologies and Politics

Technologies and politics have been central to early debates about governmentalities. As elaborated earlier a focus on technologies for governing was a pivotal element of the analytic approach outlined in Rose and Miller's early work (1992). Early sympathetic critiques of studies of governmentality identified a relative lack of attention to politics, including action orientated towards influencing the governance of political organization (Hindess, 1997: 261; O'Malley, Weir, & Shearing, 1997).

Rob Shields's chapter on nanopower and nanopolitics investigates the relationship between science and neoliberalism by closely examining the particular case of the National Institute for Nanotechnology (NINT) at the University of Alberta in Edmonton, Canada. Drawing on and extending Dean's (2010) four axes of governance by adding a fifth affective dimension, and using ethnographic and geographical approaches, Shields seeks to capture the complexity of a local place. He argues that the space in which governance occurs matters, highlighting the ways that "space images" become entangled with efforts to govern nanotechnology. In addition, Shields argues that any analysis of regimes of practices or governmentality must consider the affective dimension of governance, because otherwise we ignore the important way that individuals are governed through sentiment and mood.

In his chapter on statistics and neoliberalism Cosmo Howard engages with governmentality literature on calculative practices and quantification. He argues that this literature has overwhelmingly focused on cases where quantitative calculations have been introduced into fields where governance practices have historically been dominated by qualitative judgments. Together, these studies have the effect of suggesting a necessary relationship between an intensification of quantification and the dominance of neoliberal rationalities. Howard challenges this narrative by deliberately choosing a domain of governance that historically has been dominated by quantification, namely, *official statistics*, and asking: how are these statistical agencies changing under the influence of neoliberal rationalities?

Akin Akinwumi and Nicholas Blomley's chapter interrogates the degree to which understandings of neoliberal governmentality can illuminate reconciliation practices and politics of reconciliation in the British Columbia treaty process. These authors position land as an ethnographic object and explore the analytical and methodological value of seeing reconciliation as governmentality. If land is such an object, "ethnography" becomes about closely attending to "concrete manifestations of government" that tie land and human subjects together. This chapter asserts that reconciliation is shaped by specific governmental rationalities for problematizing and intervening in "the landed areas of social life." Here land is enlisted in strategies of governing that matter for actors tethered to the state.

Neoliberal Cities? Police and Ad Hoc Governance

The city has often been viewed as a key locus or site of the ascendency of neoliberalism, spawning concepts such as "urban neoliberalism." But this section raises doubts about neoliberalism as a key explanation of current forms and transformations in neglected areas of urban governance that operate beyond municipal governments and councils, perhaps raising doubts more than any other section.

Randy K. Lippert explores neoliberal governmentalities within an urban governance form that predates the rapid spread of neoliberal thinking and practices in North America: the condominium. He argues that these forms illustrate how urban neoliberalization often entails enlisting and modifying elements already in place rather than relying on novel invention. Although the condominium is often deemed to be obviously neoliberal in character, this governance form predates the arrival of neoliberalism and in vital ways diverges from its logic. As Lippert explains, this approach to governance is also informed by the rationality of the urban "police," most famously discussed in Foucault's writings. The chapter argues that to fully understand neoliberal governmentalities the history and current functioning of the governance assemblages with which neoliberalism has come to be closely associated need to be revisited and studied through ethnographies and related strategies aimed at unearthing the presence of other logics of government.

Mariana Valverde's chapter asserts that public-private partnerships and special-purpose agencies that provide local (usually city) infrastructure and services are typically deemed neoliberal tools. While Valverde is the only scholar in the volume who does not explicitly embrace ethnography in some manner, she does engage with neoliberalism in relation to an assemblage, thus making it clear that hers is a viable methodological alternative. Valverde argues that while local governance uses legal forms from the private domain that have neoliberal overtones, attention to historical work reveals similar hybrid governance assemblages, and for a long time one-off special agencies or authorities have taken on similar responsibilities in England and in North America. She argues that scholars seeking to understand local government should avoid trying to discern the boundary between public and private and instead should consider historical research that shows "a dizzying array of local authorities" that do not pay much heed to these boundaries. This chapter thus serves as corrective to some ethnographic work that would make these assumptions at the outset or presume that ethnography is the only means of exploring and raising doubt about the totalizing character of neoliberalism.

Neoliberal Welfare and Philanthropy

The final section explores welfare, philanthropy, and education, topics that were the focus of the earliest studies of governmentalities (Burchell et al., 1991; Hunter, 1994). The chapters within this section push the governmentality literature on welfare and education into new and diverse directions. Both contributors seek to combine an analytics of governmentality with political economy approaches, a methodology that has been most recently promoted by Fairbanks (2012), and all the contributors draw on ethnographic studies (broadly understood).

In their chapter on philanthropic interventions, education reforms, and neoliberalism, Katharyne Mitchell and Chris Lizotte examine the role that actors on the ground play in the development and transformation of neoliberal practices and rationalities and, more specifically, on the trialling of education reforms funded by large philanthropic organizations. Focusing on the specific case of a partnership between the Bill and Melinda Gates Foundation and Seattle Public Schools and drawing on documentary research, interviews, and their own on-the-ground knowledge of the case, Mitchell and Lizotte reveal the resistance these reforms encountered on the ground. Further, these "failures" incited the foundation to develop new strategies for future reform initiatives. This chapter not only highlights important changes in the governance of education within the United States, but also provides a compelling example of a developing line of focus with studies of governmentality, which are studies that examine on-the-ground responses to programs or initiatives that attempt to govern conduct in a neoliberal way.

The subjectivities, actions, and discourses of actors on the ground are also the focus of Greg Marston's study of fringe lending. Marston argues that, within studies of governmentalities and the broader literature on welfare state reform, very little attention has to date been paid to private businesses that play an important role in poverty management. This chapter brings political economy and governmentality approaches to analysis together and then holds them in tension in order to reveal the factors contributing to the growth of payday lending and the subjectivities and technologies through which the customers of payday lending are governed. Marston concludes that, when accessing high-cost credit via payday, borrowers find themselves negotiating competing discourses of paternalism and free choice.

NOTES

1 Parts of this chapter were first published in a 2014 issue of *Foucault Studies* (*18*: 11–33).
2 We are aware that some scholars who have advocated combining ethnographic methods and a post-Foucauldian governmentality approach have suggested embracing a realist governmentality approach (i.e., McKee, 2009), but this is not the approach taken in this volume.
3 A number of authors have also quantified the rise of the term "neoliberalism." Peck, Theodore, and Brenner find that of the 2,500 English-language articles that list neoliberalism as a key word, 86 per cent were published after 1998 (2010: 97), that its use exploded from a handful of articles per year in the 1980s to almost 1,000 per year between 2002 and 2005 (Boas & Gans-Morse, 2009: 138), that a third of cultural anthropology articles in the same period used "neoliberalism" (Kipnis, 2007: 383), and that there was a ninefold increase in use of the term in the Google book collection between 1990 and 2009 (Flew, 2014: 49–50).
4 The latter include the 1978 lecture on "governmentality" and the interview "Questions of method," both published in the journal *Ideology and Consciousness* (Foucault, 1979, 1981) and later published in the *Foucault Effect* (Burchell et al., 1991) together with his lecture "Politics and the study of discourse," a selection of his interviews and short writings edited by Colin Gordon (Foucault, 1980a), and later some of the essays, interviews, and Collège de France course summaries published in the three-volume series *Essential Works of Foucault* (Foucault, 2000a, 2000b, 2002).
5 See also Dean (1999: 31). It should be noted that Rose does not specifically mention archival methods, and in Rose et al. he and his collaborators state that the tools of governmentality are "compatible with many other methods" (2006: 101).
6 Collier makes a somewhat similar argument but takes inspiration from Actor-Network Theory (ANT), arguing that he refuses to take figures such as society as pre-given categories but instead views them as assemblages and insists that "construction has to be accounted for" (2011b: 27).

REFERENCES

Barry, A., Osborne, T., & Rose, N. (1996a). Introduction. In A. Barry, T. Osborne, & N. Rose (Eds), *Foucault and political reason: Liberalism, neo-liberalism, and rationalities of government* (1–17). Chicago: University of Chicago Press.

Barry, A., Osborne, T., & Rose, N. (1996b). *Foucault and political reason: Liberalism, neo-liberalism, and rationalities of government*. Chicago: University of Chicago Press.

Besley, T. (2002). *Counseling youth: Foucault, power, and the ethics of subjectivity*. Westport, CT: Praeger.

Binkley, S. (2009). The work of neoliberal governmentality: Temporality and ethical substance in the tale of two dads. *Foucault Studies, 6*, 60–78.

Boas, T.C., & Gans-Morse, J. (2009). Neoliberalism: From new liberal philosophy to anti-liberal slogan. [SCID]. *Studies in Comparative International Development, 44*(2), 137–61. http://dx.doi.org/10.1007/s12116-009-9040-5

Brenner, N., & Theodore, N. (2002). Cities and the geographies of "actually existing neoliberalism." *Antipode, 34*(3), 349–79. http://dx.doi.org/10.1111/1467-8330.00246

Brenner, N., Peck, J., & Theodore, N. (2010). Variegated neoliberalization: Geographies, modalities, pathways. *Global Networks, 10*(2), 182–222. http://dx.doi.org/10.1111/j.1471-0374.2009.00277.x

Bröckling, U., Krasmann, S., & Lemke, T. (2011). *Governmentality: Current issues and future challenges*. New York: Routledge.

Burchell, G., Gordon, C., & Miller, P. (1991). *The Foucault effect: Studies in governmentality. With two lectures by and an interview with Michel Foucault.* Chicago: University of Chicago Press. http://dx.doi.org/10.7208/chicago/9780226028811.001.0001

Caswell, D., Marston, G., & Larsen, J.E. (2010). Unemployed citizen or "at risk'" client? Classification systems and employment services in Denmark and Australia. *Critical Social Policy, 30*(3), 384–404. http://dx.doi.org/10.1177/0261018310367674

Collier, S.J. (2009). Topologies of power: Foucault's analysis of political government beyond "governmentality." *Theory, Culture & Society, 26*(6), 78–108. http://dx.doi.org/10.1177/0263276409347694

Collier, S.J. (2011a). *Foucault, assemblages, and topology: Interview with Simon Dawes*. Available at http://theoryculturesociety.org/interview-with-stephen-j-collier-on-foucault-assemblages-and-topology/. Accessed 21 May 2015.

Collier, S.J. (2011b). *Post-Soviet social: Neoliberalism, social modernity, biopolitics*. Princeton: Princeton University Press. http://dx.doi.org/10.1515/9781400840427

Collier, S.J. (2012). Neoliberalism as big leviathan, or …? A response to Wacquant and Hilgers. *Social Anthropology, 20*(2), 186–95. http://dx.doi. org/10.1111/j.1469-8676.2012.00195.x

Collier, S.J. (2013). "Fieldwork as technique for generating what kind of surprise? Thoughts on post-Soviet social in light of 'fieldwork/research.'" Talk given at University of California, Irving. Available at: https:// stephenjcollier.com/talks/.

Dean, M. (1994). *Critical and effective histories: Foucault's methods and historical sociology.* London, New York: Routledge. http://dx.doi. org/10.4324/9780203414217

Dean, M. (1998). Administering asceticism: Re-working the ethical life of the unemployed citizen. In M. Dean & B. Hindess (Eds), *Governing Australia: Studies in contemporary rationalities of government* (87–107). Melbourne: Cambridge University Press.

Dean, M. (1999). *Governmentality: Power and rule in modern society.* London, Thousand Oaks, CA: Sage.

Dean, M. (2010). *Governmentality: Power and rule in modern society.* 2nd ed. Los Angeles: Sage.

Deleuze, G., & Guattari, F. (1987). *A thousand plateaus: Capitalism and schizophrenia.* Minneapolis: University of Minnesota.

Donzelot, J. (1979). The poverty of political culture. *Ideology & consciousness, 5*(1), 73–86.

Donzelot, J., & Gordon, C. (2009). Governing liberal societies: The Foucault effect in the English-speaking world. In M.A. Peters, A.C. Besley, M. Olssen, S. Maurer, & S. Weber (Eds), *Governmentality studies in education* (3–16). Rotterdam: Sense.

Fairbanks, R.P. (2012). On theory and method: Critical ethnographic approaches to urban regulatory restructuring. *Urban Geography, 33*(4), 545–65. http://dx.doi.org/10.2747/0272-3638.33.4.545

Fejes, A. (2008). To be one's own confessor: Educational guidance and governmentality. *British Journal of Sociology of Education, 29*(6), 653–64. http://dx.doi.org/10.1080/01425690802423320

Ferguson, J. (2006). *Global shadows: Africa in the neoliberal world order.* Durham, NC: Duke University Press. http://dx.doi.org/10.1215/9780822387640

Flew, T. (2012). Michel Foucault's The Birth of Biopolitics and contemporary neoliberalism debates. *Thesis Eleven, 108*(1), 44–65. http://dx.doi. org/10.1177/0725513611421481

Flew, T. (2014). Six theories of neoliberalism. *Thesis Eleven, 122*(1), 49–71. http://dx.doi.org/10.1177/0725513614535965

Flint, J. (2003). Housing and ethopolitics: Constructing identities of active consumption and responsible community. *Economy and Society, 32*(4), 611–29. http://dx.doi.org/10.1080/0308514032000107628

Forsey, M.G. (2010). Ethnography as participant listening. *Ethnography, 11*(4), 558–72. http://dx.doi.org/10.1177/1466138110372587

Foucault, M. (1979). Governmentality. *Ideology & Consciousness, 6*(1), 5–21.

Foucault, M. (1980a). *Power/knowledge: Selected interviews and other writings, 1972–1977.* Edited by C. Gordon. New York: Pantheon Books.

Foucault, M. (1980b). The confession of the flesh. In C. Gordon (Ed.), *Power/ knowledge: Selected interviews and other writings, 1972–1977* (194–228). New York: Pantheon Books.

Foucault, M. (1981). Questions of method. *Ideology & Consciousness, 8*(1), 3–14.

Foucault, M. (1984). On the genealogy of ethics: An overview of work in progress. In P. Rabinow (Ed.), *The Foucault reader* (340–72). New York: Pantheon Books.

Foucault, M. (1987). About the beginnings of the hermeneutics of the self. In J.R. Carrette (Ed.), *Religion and culture* (158–81). Manchester, UK: Manchester University Press.

Foucault, M. (1988). The ethic of care of the self as a practice of freedom: An interview. In J.W. Bernauer & D.M. Rasmussen (Eds), *The final Foucault* (1–20). Cambridge, MA: MIT Press.

Foucault, M. (2000a). *Ethics: Essential works of Foucault, 1954–1984.* Vol. 1. Victoria, AUS: Penguin.

Foucault, M. (2000b). *Aesthetics: Essential works of Foucault, 1954–1984.* Vol. 1. Victoria, AUS: Penguin.

Foucault, M. (2000c). Questions of method. In J.D. Faubion (Ed.), *Michel Foucault power* (223–38). New York: New Press.

Foucault, M. (2002). *Power: Essential works of Foucault, 1954–1984.* Vol. 3. Victoria, AUS: Penguin.

Foucault, M. (2008). *The birth of biopolitics.* New York: Palgrave Macmillan. http://dx.doi.org/10.1057/9780230594180

Fridman, D. (2010). A new mentality for a new economy: performing the homo economicus in Argentina (1976–83). *Economy and Society, 39*(2), 271–302. http://dx.doi.org/10.1080/03085141003620170

Gordon, C. (1987). The soul of the citizen: Max Weber and Michel Foucault on rationality and government. In S. Lash & S. Whimster (Eds), *Max Weber: Rationality and modernity* (293–316). London: Allen & Unwin.

Gordon, C. (1991). Governmental rationality: An introduction. In G. Burchell, C. Gordon, & P. Miller (Eds), *The Foucault effect: Studies in governmentality* (1–52). Chicago: University of Chicago Press.

Haggerty, K.D., & Ericson, R.V. (2000). The surveillant assemblage. *British Journal of Sociology, 51*(4), 605–22. http://dx.doi.org/10.1080/00071310020015280

Hamann, T.H. (2009). Neoliberalism, governmentality, and ethics. *Foucault Studies, 6*(1), 37–59.

Harris, P. (2001). From relief to mutual obligation: Welfare rationalities and unemployment in 20th-century Australia. *Journal of Sociology (Melbourne, Vic.), 37*(1), 5–26. http://dx.doi.org/10.1177/144078301128756175

Harvey, D. (2007). *A brief history of neoliberalism.* New York: Oxford University Press.

Higgins, V., & Larner, W. (Eds). (2010). *Calculating the social: Standards and the reconfiguration of governing.* Basingstoke, UK, New York: Palgrave Macmillan. http://dx.doi.org/10.1057/9780230289673

Hilgers, M. (2010). The three anthropological approaches to neoliberalism. *International Social Science Journal, 61*(202), 351–64. http://dx.doi.org/10.1111/j.1468-2451.2011.01776.x

Hindess, B. (1997). Politics and governmentality. *Economy and Society, 26*(2), 257–72. http://dx.doi.org/10.1080/03085149700000014

Hoffman, L.M. (2010). *Patriotic professionalism in urban China: Fostering talent.* Philadelphia, PA: Temple University Press.

Hunter, I. (1994). *Rethinking the school: Subjectivity, bureaucracy, criticism.* St Leonards, NSW: Allen & Unwin.

Keil, R. (2011). The global city comes home: Internalized globalization in Frankfurt Rhine-Main. *Urban Studies (Edinburgh, Scotland), 48*(12), 2495–517. http://dx.doi.org/10.1177/0042098011411946

King, D.S. (1987). *The new right: Politics, markets and citizenship.* Homewood, IL: Dorsey Press. http://dx.doi.org/10.1007/978-1-349-18864-2

Kipnis, A. (2007). Neoliberalism reified: Suzhi discourse and tropes of neoliberalism in the People's Republic of China. *Journal of the Royal Anthropological Institute, 13*(2), 383–400. http://dx.doi.org/10.1111/j.1467-9655.2007.00432.x

Larner, W. (2000). Neoliberalism: Policy, ideology, governmentality. *Studies in Political Economy, 63*(63), 5–25.

Larner, W. (2009). Neoliberalism, Mike Moore, and the WTO. *Environment & Planning A, 41*(7), 1576–93. http://dx.doi.org/10.1068/a41142

Larner, W., & Walters, W. (2004). *Global governmentality: Governing international spaces.* New York: Routledge.

Legg, S. (2011). Assemblage/apparatus: Using Deleuze and Foucault. *Area*, 43(2), 128–33. http://dx.doi.org/10.1111/j.1475-4762.2011.01010.x

Lemke, T. (2001). "The birth of biopolitics": Michel Foucault's lecture at the Collège de France on neoliberal governmentality. *Economy and Society*, 30(2), 190–207. http://dx.doi.org/10.1080/03085140120042271

Lemke, T. (2002). Foucault, governmentality, and critique. *Rethinking Marxism*, 14(3), 49–64. http://dx.doi.org/10.1080/089356902101242288

Levitas, R. (Ed.) (1986). *The ideology of the new right*. Cambridge: Polity Press.

Li, T.M. (2007a). *The will to improve: Governmentality, development, and the practice of politics*. Durham, NC: Duke University Press. http://dx.doi.org/10.1215/9780822389781

Li, T.M. (2007b). Practices of assemblage and community forest management. *Economy and Society*, 36(2), 263–93. http://dx.doi.org/10.1080/03085140701254308

Lippert, R. (1998). Rationalities and refugee resettlement. *Economy and Society*, 27(4), 380–406. http://dx.doi.org/10.1080/03085149800000026

Lippert, R. (1999). The relevance of governmentality to understanding the international refugee regime. *Alternatives: Social Transformation and Humane Governance*, 24(3), 295–328. http://dx.doi.org/10.1177/030437549902400302

Lippert, R. (2010). Mundane and mutant devices of power: Business improvement districts and sanctuaries. *European Journal of Cultural Studies*, 13(4), 477–94. http://dx.doi.org/10.1177/1367549410377155

Lippert, R. (2014). Neo-liberalism, police, and the governance of little urban things. *Foucault Studies*, 18(1), 49–65.

Lippert, R., & O'Connor, D. (2003). Security assemblages: Airport security, flexible work, and liberal governance. *Alternatives*, 28(3), 331–58. http://dx.doi.org/10.1177/030437540302800302

Lippert, R.K. (2005). *Sanctuary, sovereignty, sacrifice: Canadian sanctuary incidents, power, and law*. Vancouver: UBC Press.

Lippert, R.K. (2009). Signs of the surveillant assemblage: Privacy regulation, urban CCTV, and governmentality. *Social & Legal Studies*, 18(4), 505–22. http://dx.doi.org/10.1177/0964663909345096

Lippert, R.K., & Pyykkönen, M. (2012a). Introduction. Immigration, governmentality, and integration assemblages. *Nordic Journal of Migration Research*, 2(1), 1–4. http://dx.doi.org/10.2478/v10202-011-0021-1

Lippert, R.K., & Pyykkönen, M. (2012b). Contesting family in Finnish and Canadian immigration and refugee policy. *Nordic Journal of Migration Research*, 2(1), 45–56. http://dx.doi.org/10.2478/v10202-011-0026-9

Lippert, R.K., & Stenson, K. (2010). Advancing governmentality studies: Lessons from social constructionism. *Theoretical Criminology*, 14(4), 473–94. http://dx.doi.org/10.1177/1362480610369328

McKee, K. (2009). Post-Foucauldian governmentality: What does it offer critical social policy analysis? *Critical Social Policy, 29*(3), 465–86. http://dx.doi.org/10.1177/0261018309105180

Miller, P., & Rose, N. (1990). Governing economic life. *Economy and Society, 19*(1), 1–31. http://dx.doi.org/10.1080/03085149000000001

Mitchell, K. (2006). Neoliberal governmentality in the European Union: Education, training, and technologies of citizenship. *Environment and Planning. D, Society & Space, 24*(3), 389–407. http://dx.doi.org/10.1068/d1804

Mudge, S.L. (2008). What is neoliberalism? *Socio-economic Review, 6*(4), 703–31. http://dx.doi.org/10.1093/ser/mwn016

O'Malley, P. (1992). Risk, power and crime prevention. *International Journal of Human Resource Management, 21*(3), 252–75.

O'Malley, P., & Valverde, M. (2014). Foucault, criminal law, and the governmentalization of the state. In M.D. Dibber (Ed.), *Foundational texts in modern criminal law* (317–33). Oxford: Oxford University Press. http://dx.doi.org/10.1093/acprof:oso/9780199673612.003.0017

O'Malley, P., Weir, L., & Shearing, C. (1997). Governmentality, criticism, politics. *Economy and Society, 26*(4), 501–17. http://dx.doi.org/10.1080/03085149700000026

Ong, A., & Collier, S.J. (2005). *Global assemblages: Technology, politics, and ethics as anthropological problems*. Malden, MA: Blackwell.

Peck, J. (2013). Explaining (with) neoliberalism. *Territory, Politics. Governance: An International Journal of Policy, Administration and Institutions, 1*(2), 132–57.

Peck, J., & Theodore, N. (2012). Reanimating neoliberalism: Process geographies of neoliberalization. *Social Anthropology, 20*(2), 177–85. http://dx.doi.org/10.1111/j.1469-8676.2012.00194.x

Peck, J., Theodore, N., & Brenner, N. (2010). Postneoliberalism and its malcontents. *Antipode, 41*(s1), 94–116. http://dx.doi.org/10.1111/j.1467-8330.2009.00718.x

Pusey, M. (1989). *Economic rationalism in Canberra: A nation-building state changes its mind*. Cambridge: Cambridge University Press. http://dx.doi.org/10.1017/CBO9780511597121

Rabinow, P. (2003). *Anthropos today: Reflections on modern equipment*. Princeton: Princeton University Press.

Rose, N. (1996). Governing "advanced" liberal democracies. In A. Barry, T. Osborne, & N. Rose (Eds), *Foucault and political reason: Liberalism, neoliberalism and rationalities of government* (37–64). Chicago: University Of Chicago Press.

Rose, N. (1999). *Powers of freedom: Reframing political thought*. New York: Cambridge University Press. http://dx.doi.org/10.1017/CBO9780511488856

Rose, N. (2000). Community, citizenship, and the third way. *American Behavioral Scientist*, 43(9), 1395–411. http://dx.doi.org/10.1177/00027640021955955

Rose, N., & Miller, P. (1992). Political power beyond the state: Problematics of government. *British Journal of Sociology*, 43(2), 173–205. http://dx.doi.org/10.2307/591464

Rose, N., O'Malley, P., & Valverde, M. (2006). Governmentality. *Annual Review of Law and Social Science*, 2(1), 83–104. http://dx.doi.org/10.1146/annurev.lawsocsci.2.081805.105900

Simon, J. (2007). *Governing through crime: How the war on crime transformed American democracy and created a culture of fear*. New York: Oxford University Press.

Steger, M.B., & Roy, R.K. (2010). *Neoliberalism: A very short introduction*. New York: Oxford University Press. http://dx.doi.org/10.1093/actrade/9780199560516.001.0001

Stenson, K., & Watt, P. (1999). Governmentality and "the death of the social"? A discourse analysis of local government texts in south-east England. *Urban Studies (Edinburgh, Scotland)*, 36(1), 189–201. http://dx.doi.org/10.1080/0042098993817

van Apeldoorn, B., & De Graaff, N. (2012). Beyond neoliberal imperialism? The crisis of American empire. In H. Overbeek & B. van Apeldoorn (Eds), *Neoliberalism in crisis* (207–28). New York: Palgrave Macmillan. http://dx.doi.org/10.1057/9781137002471.0017

Wacquant, L. (2012). Three steps to a historical anthropology of actually existing neoliberalism. *Social Anthropology*, 20(1), 66–79. http://dx.doi.org/10.1111/j.1469-8676.2011.00189.x

Walters, W. (1997). The "active society": New designs for social policy. *Policy and Politics*, 25(3), 221–34. http://dx.doi.org/10.1332/030557397782453264

Wilson, D. (2004). Toward a contingent urban neoliberalism. *Urban Geography*, 25(8), 771–83. http://dx.doi.org/10.2747/0272-3638.25.8.771

Yeatman, A. (1990). *Bureaucrats, technocrats, femocrats: Essays on the contemporary Australian state*. Sydney: Allen & Unwin.

PART 1

Neoliberal Subjectivities? Examining Resilience, Investment, and Enterprise

2 Creating Resilient Subjects: The Coexist Project

WENDY LARNER AND SIMON MORETON

"Resilience" has become a new lingua franca for how we should live, mobilized in realms as diverse as climate change, global urbaniza- tion, organizational change, community building, and personal development. In its broadest sense, it signifies the nature of living in a context understood to be characterized by hitherto unprec- edented levels of risk and uncertainty in multiple arenas (Cooper, 2010; O'Malley, 2010). Whereas growing numbers of social scientists attribute the increasing prominence of the term to the discourses and practices of neoliberalism (see Evans, 2011; Joseph, 2013), this chap- ter suggests that the political genealogies and geographies of resil- ience may be more complex. Of particular interest is how the term is being tactically used by grass-roots political subjects who are actively experimenting with relational ways of living. Drawing on ongoing empirical research conducted with Coexist, an umbrella organization set up in 2008 to support environmental and community groups in Bristol, UK, this chapter argues that resilience might be better under- stood as an example of an "after neoliberal" (Larner, Le Heron, & Lewis, 2007) form of subjectification. While resilience is undoubtedly shaped by the legacies of neoliberalism, it is becoming central to the ways in which communities and individuals are envisioning and experimenting with alternative environmental, economic, political, and social futures.

The chapter begins by briefly reviewing the current social scien- tific debates about resilience to position this claim. The focus is on the recent uptake of the term "resilience" by grass-roots activists and com- munity organizations who are challenging prevailing environmental, economic, political and social norms (see also Cretney & Bond, 2014).

Drawing from wider accounts of neoliberalism as governmentality, the chapter shows how the new emphasis on resilience involves attempts to foster particular kinds of organizational and individual behaviour. While much of the literature on contemporary forms of subjectification has concentrated on the processes of marketization and *homo economicus* (Larner, 2012; Li in this volume; Rankin, 2001), attempts to "conduct the conduct" of neoliberal subjects have always involved diverse political imperatives and differentiated experiences (Hoffman, 2010; Larner, 2006; Ong, 1999; Pykett, Jones, Whitehead, Huxley, Strauss, et al., 2011). Such studies suggest that understanding the cultivation and fostering of subjectivities may require a more careful focus on the multiple processes, practices, and power relations that shape our political present. Resilience is one such arena.

The remainder of the chapter illustrates this claim through the empirical case, developing an argument that Coexist exemplifies the new emphasis on resilience as the way we should live after neoliberalism. Although this political formulation rejects market-led and individualized models often associated with neoliberalism, it is not a return to the state-led formulations of Keynesian welfarism and developmentalism either. As the influential UK think-tank Demos (2009: 1) states, "next generation resilience relies on citizens and communities, not the institutions of the state." In this "after neoliberal" political formation there is explicit recognition of the interconnectedness of environmental, economic and social processes. The political imperatives of resilience thinking privileges modes of experimental governance in which policy makers, researchers, businesses, and communities are collectively charged with finding new paths to sustainability (Evans, 2011).

In drawing on this "resilience thinking" organizations such as Coexist are positioning themselves as vehicles for wider forms of transformative change. Close examination of such initiatives allows us to trace the features associated with resilience as a new form of politics and governance, identifying the organizational forms resilience takes; the resources, capacities, and skills it mobilizes; and how resilient subjects are imagined and constituted. In identifying these features we also underline the wider argument that grass-roots and alternative projects, such as anti-poverty, environmental, and creative groups, often can "prefigure" new governmental forms in ways that have not been fully recognized by the existing literature on neoliberalism (Larner, 2000, 2014; Larner & Craig, 2005; Lorey, 2011; Newman, 2012) and that these groups have crucial implications for how developments after neoliberalism might be understood and shaped.

While not strictly ethnographic in the traditional anthropological sense, the chapter draws on research conducted as part of an ongoing collaboration between the authors and key members of Coexist. In 2010, when the embryonic organization was still working out its overall approach and position in wider Bristol-based networks, Coexist approached a member of staff at the University of Bristol to see if there was any interest in joint activity. In the four years since the initial approach, combined projects have included a formal occupant survey (run between December 2010 and January 2011), a four-month period of ethnographic fieldwork and participant observation carried out by Moreton (2011), and a student-led use survey (2012). Most recently, Coexist has become part of "Productive Margins," a wider funded program that brings together a range of Bristol-based community organizations and academic researchers to co-produce research projects on the topic of regulatory reform (2013–18). This chapter draws on data generated in the early years of this collaborative research program, which aimed to formally document Coexist activities, capturing what was actually happening in Hamilton House: the experiences of those involved, their working practices, and their understandings of the Coexist project.

"Resilience" after Neoliberalism

Resilience has an established, common sense meaning related to the strategies of endurance that people use to facilitate their day-to-day living. More recently, however, resilience has become a ubiquitous term for how we should live in this historical moment. Whereas sustainability was the dominant rubric for environmental policy in the 1990s, as its successor resilience has a much broader remit that draws together environmental, economic, social, and psychic processes. In social science literature the term has begun to proliferate, appearing in diverse empirical fields, including environmental geography (Bulkeley, Broto, & Edwards, 2012; Folke, 2006; Pelling, 2010; Weichselgartner & Kelman, 2014), cities (Evans, 2011; Stumpp, 2013), urban planning (Eraydin & Taşan-Kok, 2013; Raco & Street, 2012; Wagenaar & Wilkinson, 2015), and economic geography (Christopherson, Michie, & Tyler, 2010; Berkes & Ross, 2013; Bristow & Healy, 2014; Martin, 2012). Perhaps even more important, resilience now features strongly in the "grey literature" of think-tanks, environmental consultancies, social movements (Bristow, 2010: 154), as well as various funding programs (Welsh, 2014). The rapidly growing literature, which now includes at least two eponymous journals, has given rise to a variety of attempts to create taxonomies and definitions of resilience (see Brown, 2014). This chapter does not even attempt to rehearse these debates. Indeed, it is the very ubiquity and slipperiness of the term that is of interest.

Three interconnected sets of explanations can be mobilized to explain why resilience has become such a ubiquitous term over the last decade. The first set of explanations might be seen as socio-structural. It is widely accepted that the world has become even more interconnected, but also a much less certain place – environmentally, economically, geopolitically, and socially. After 9/11 increased recognition of global warming and the advent of a sustained global financial crisis, among other developments, established forms of power are no longer taken for granted, problems are no longer given, accepted forms of expertise are no longer accepted as adequate, and our futures are less self-evident (Larner, 2011). For some it is these intersecting crises and the enhanced perception of vulnerability that explains the search for new paths to resilience (Christopherson et al., 2010; Eraydin, 2013; Shaw & Maythorne, 2013). Recent debates about the growth of precarity also capture this new sense of unpredictability and vulnerability and have similarly given rise to claims for the need for resilience. For example, Ettlinger (2011: 320) builds on observed links between the growing uncertainty of labour regimes and terrorism and draws on Judith Butler to argue that precarity is a more generalizable condition of vulnerability relative to contingency and the inability to predict. She argues that resilience based on multiple networked dynamics and cooperative forms of politics participating in, rather than opposing, governance is what is needed in this unpredictable world.

The second set of explanations for the rise of the term resilience is governmental. Often drawing on Foucauldian accounts, commentators in diverse fields, including international relations, security studies, urban geography, science and technology studies, criminology, and social work, have identified the new prominence of uncertainty, risk, and resilience as political technologies. There are conflicting explanations for this new centring of resilience, but commentators are agreed that this approach presents the world as being in constant flux and periodic crisis; understands that change is inevitable; and presupposes that the capacities of individuals, communities, and cities need to be developed so they can adapt to these changes. In important contributions to this debate, Walker & Cooper (2011) and O'Malley (2010: 448) elaborate these claims, showing that resilience has joined risk and preparedness as part of the security assemblage emerging around "uncertain and potentially traumatic futures." While the genealogical research of these authors takes them in different directions, both show that routines and resources for emergencies have moved from those that are precisely calculable (risk) to those that are imaginable (preparedness) and are now more encompassing of unimaginable and uncertain futures (resilience). Most important, this new discourse of resilience has facilitated a shift in understandings of human security that replace state-centric approaches with those that encourage development of community institutions (see Dorow in this volume) and citizen groups (Chandler, 2012; Coaffee, 2013).

The third arena in which the concept of resilience is of growing import, and of central concern to this chapter, is subjectification. Geographers are familiar with the well-rehearsed discussion of neoliberal subjectivities in which contractualism and calculative practices foster responsibilization. But resilient subjects are not the entrepreneurialized rational subjects of economic government (Rose, 1999), nor are they the geneticized subjects imagined by neuroscience and behavioural economics (Berndt, 2014; Pykett, 2013). This chapter will show that resilient subjects are self-sustaining, empowered, collectivized subjects who envisage "real utopias" (Fung & Wright, 2003) and engage in prefigurative politics. In doing so, they create alternative environmental, economic, and social futures to those commonly associated with global competitiveness and untrammelled economic growth (Bristow, 2010). For example, in her reflections on the global environmental crisis and reported arrival of the Anthropocene, Gibson-Graham (2011: 4) asks us "to recognize that we are all participants in a 'becoming world' in which everything is interconnected and learning happens in a stumbling, trial and error kind of way." She argues for what she calls "new adventures in living" that trigger the self-organizing resilient local economies and empowered subjects that can underpin a new mode of humanity. Similarly, Swyngedouw (2009) has seen resilience as a potential basis for an environmental politics of progressive social movements.

Resilient Subjectivities

So how should we understand this new way of being in the world? The most common answer to this question is that resilience represents a further extension of the individualizing, autotomizing aspects of neoliberalism. For example, Reid (2012) argues that in its emphasis on entrepreneurial practices of subjectivity the new focus on resilience resonates with neoliberal discourses of capitalism. He argues that "resilient" peoples do not look to states to secure their well-being because they have been disciplined into believing in the necessity of securing it for themselves (Reid, 2012: 69). In a thought-provoking intervention, Duffield (2011: 14) argues resilience is an ideology attuned to the uncertainties of a neoliberal economy and an abandonment of political subjectivity. He suggests that a resilient life is able to exist on the edge of survivability and is adapted to uncertainty and surprise. In a related intervention, Chandler (2010: 295) speculates that resilience might be a discourse of Western self-limitation that reflects the impossibility of control in a context where the goal of creating resilient societies and subjects cannot be ensured by a global sovereign. In this context resilience becomes a means of democratizing, developing, and securing the "Other" under the imperative of becoming safe for both itself and others. In discussing these issues, Vinay Gidwani suggested that the new emphasis on resilience may have arisen out of the failures of both structural adjustment and millennium development goals. Reflecting on Li's (2010) work (see also Li in this volume), he argued that we no longer see the "will to improve" but rather "let die" as a way of coping with the growing phenomenon of surplus humanity. Resilience is the fate that faces the rest of us.

Certainly, the aspiration of resilience is no longer confined to marginal places and people. Moreover, it is precisely the polysemic nature of the term resilience and the existence of these debates over its significance that should alert us to the idea that resilience is usefully seen as a governmental assemblage (see Li in this volume) that has been produced by evolving and intersecting forms of politics. In the field of environment, for example, there were early accounts of links between social and ecological resilience (Adger, 2000). In regional development these debates about environmental adaptation became linked to the fields of evolutionary economics and evolutionary economic geography (Bristow & Healy, 2014; Christopherson et al., 2010; Martin, 2012). In their account Walker and Cooper (2011) attribute the rise of resilience thinking to the transition from ecological theory to a socioecological governance framework that mirrors Hayekian complexity-based economics. In contrast O'Malley (2010), working in the field of criminology, attributes the rise of resilience to an unexpected articulation of liberalism, medicine, and militarism. Moreover, he observes but does not explore the fact that theories of resilience can also be found in social work and self-help discourses, areas in which feminism has had a huge influence. While developing a critical genealogy of the term is not the task here, these contrasting accounts suggest that the genealogy of resilience may be more heterogeneous than we have previously understood.

Despite their differing explanations of the rise of resilience thinking, these commentators agree that resilience signifies the nature of living in a context of unprecedented levels of risk and uncertainty in multiple arenas and unknown amounts of qualitative change, all of which are both presupposed and normalized. As Lentzos and Rose (2009: 243) observe in what has become a seminal contribution to this debate, "Resilience implies a systematic, widespread, organizational, structural and personal strengthening of subjective and material arrangements so as to be better able to anticipate and tolerate disturbances in complex worlds without collapse, to withstand shocks, and to rebuild as necessary." Lentzos and Rose are clear that this formulation of resilience is the opposite of the "Big Brother" state. There is no imagined sovereign who is directing and controlling society with the aspiration of resilience, but equally important, resilience is not neoliberalism as we know and currently theorize it.

More specifically, the emphasis on resilience can be seen as part of a more general move towards individual and collective empowerment that has even further displaced states, professionals, and technocrats. The new emphasis on pedagogical programs of development, modernization, and reform has already been identified and discussed in accounts of the post-welfare state (Newman, 2010). But resilience thinking almost entirely dispenses with the state and its experts in the cultivation of self-provisioning and self-management. Resilience also rests on new conceptions of leadership, courage, will-power, fortitude, and character not as natural abilities but as skills that can be learned by everyone. In this context we need to know more about the institutional and technical arrangements implanted to foster resilience and to enhance the capacities of human agents for particular kinds of action and cognition (Çalişkan & Callon, 2010). What knowledge, know-how, and skills are developed and mobilized in the process of designing and managing these new configurations? How are the behaviours and competencies required by individual and collective resilient subjects acquired? We also need to know much more about the effects of resilience. What unprecedented forms of society and economy are invented when resilient subjects that are explicitly attentive to relationships of solidarity are created? To answer these questions we now turn to the case study of Coexist.

The Coexist Project

Coexist is an umbrella organization that manages spaces in which people can "coexist" with themselves, with each other, and with the environment. They occupy a formerly derelict office building – Hamilton House – beside Banksy's famous Mild, Mild West mural of a teddy bear with a petrol bomb in Stokes Croft, Bristol, UK. The 55,000-square-foot building was once called Finance House and was populated by office workers employed in the back office of a London-based insurance company. The insurance company moved out in the mid-2000s, leaving the building vacant and increasingly run-down. For several years the area at the front of the empty building was a favoured spot for street drinkers. In 2008 Coexist took over the building, and by 2010 Hamilton House accommodated over 120 tenants, ranging from visual artists to music production companies, global development charities, and alternative energy organizations. It also had community spaces available for rent and hosted public events, including educational initiatives, community-building forums, and diverse forms of cultural activity. More recent additions include a purpose-built community kitchen, a well-being suite used for therapy and classes, and music studios, all of which are for hire at deliberately affordable rates.

Coexist was established within multiple politicized and pluralized discourses. The organization's explicit point of reference was the UK government's *Egan Review of Skills for Sustainable Communities* (Office of the Deputy Prime Minister, 2004). The *Egan Review* is framed through the language of sustainability and links environmental action, economic activity, infrastructural provision, and community engagement as a part of a mutually constituted ecosystem (see Figure 1). The Coexist vision is that they will provide the skills and resources that communities need to thrive in this ecosystem and in doing so help these communities become – to use their own word – resilient. Over time this emphasis on creating sustainable communities by fostering resilience has come to occupy a central place in the discourses and practices of those who orbit around Hamilton House.

Coexist's particular approach to sustainability poses the long-standing question of grass-roots community development in new ways. Rather than the more familiar model of community development

Figure 1: Components of sustainable communities (Office of the Deputy Prime Minister, 2004: 19)

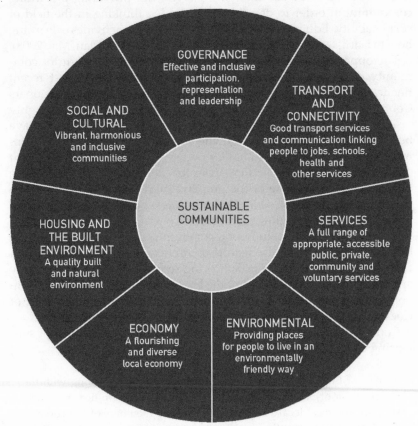

based on providing basic needs to lower-income and/or marginalized communities, their model expects community members to know their needs (as outlined in the *Egan Review* regarding food, transport, and so on [ibid.]). It also presupposes that these actors have the social and cultural capital to access Hamilton House as a way of fulfilling these needs. To make this point in more conceptual terms, Coexist envisages a "learning community" comprising "active citizens" willing and able to become whomever they like within the mutually supportive space of Hamilton House. Thus, while Hamilton House is described as a "community centre," Coexist envisages this community in non-traditional ways. To use their words: "We are providing a learning community in order to develop leading edge thinking in the field of sustainability. Education is at the heart of all our activities – learning how to learn, teaching how to teach" (Coexist Business Plan, May 2008).

The opportunity to put this vision of an alternative form of community development into practice arose following the most recent financial crisis. The owners of Hamilton House – property developers Connolly and Callaghan – had received planning permission for eighty residential apartments in the building, which subsequently would have been made available for private ownership. In this regard, the development might have followed familiar patterns of property development and gentrification (see also Lippert in this volume). These plans were curtailed by the collapse of the property market in late 2008 and the subsequent risk of a rates bill resulting from a proposed "empty building tax." The two founding members of Coexist were initially engaged to conduct a feasibility study of a community-led reoccupation of the building, which would also feature environmental initiatives such as a roof garden and solar panels. Following receipt of the feasibility study and the ensuing development of the business plan quoted above, Connolly and Callaghan subsequently seed-funded the establishment of Coexist as a Community Interest Company.

Connolly and Callaghan publicly describe their motivations for supporting this initiative in the following way:

> As part of Bristol's "Green Capital" initiative, we are assisting with the regeneration of Stokes Croft and St Paul's by offering office space and support to ethical local ventures. As founding members of the community interest company Egregoria, we are working on a number of joint initiatives that together form a centre of excellence in sustainable communities. We are also supporting the formation of a charitable venture to bring together a variety of organizations that work with communities to reduce homelessness and crime. (ConnollyandCallaghan.co.uk; accessed July 2011)

Of course, such claims can be regarded with some scepticism. The gentrification literature has identified how the long-established practices of low-income tax credits, land and/or building write-downs, and planning permissions have underpinned the development of artist's studios and live/work units. But unlike many such examples, Hamilton House is not being used by developers to increase profitable turnover in run-down neighbourhoods, as described in Zukin's (1982) "artistic mode of production." Rather, the joint aim of the developers and Coexist was to establish a community centre that would profile grass-roots environmental initiatives and play an active role in the neighbourhood itself, thereby delivering on their shared ambition for a more sustainable future.

As Hamilton House gradually filled up, Coexist's (self-named) "directors" often found themselves acting in relatively straightforward ways as building managers, overseeing the well-being of various (low-) rent-paying individuals and small group tenants. Each of the directors received a living wage and in exchange oversaw pragmatic concerns such as Health and Safety, Fire Safety, building maintenance, rent collection, problem solving, and service delivery. More broadly, however, the directors understood Coexist as a "container" project that would provide a supportive environment to help social enterprises, creative workers, and low-carbon organizations grow and flourish. One director, coming from a background in dance therapy, explains that his ambition was to establish "safety, trust and a sense of play for those involved" (Jamie Pike, Coexist founding director).

An emphasis on active and deliberate networking informed wider aspects of Coexist's work. The ambition was to create what those involved came to call a "culture of co-operation" where diverse actors can come together and use Hamilton House as a vehicle to achieve their personal and professional aims. Co-location would foster relationships that would ferment innovation, give rise to new social enterprises, and help generate the resilience they believe a sustainable future needs. A survey conducted with the building's occupants suggested that this aim was widely understood: "The sense that I get from Hamilton House is that there are a lot of projects that have been set up in the interest of the local, and broader, communities. I don't really think of it as an arts space ... but rather a space in which groups can rent affordable workshop and studio space with which they help to make a difference outside of [the building]" (Hamilton House tenant, interview 2011).

This approach to building sustainable communities influences who is offered space in Hamilton House. The ambition is to provide space for those who themselves are widely networked, thereby serving as "catalysts" who can bring new relationships to the project and increase its credibility and cultural viability. It also underpins efforts to bring outside partners into Hamilton House to work with building tenants on the assumption that new partnerships and learning opportunities will emerge from these types of interaction. This approach has included, for example, Bristol City College using spaces for art classes, non-resident groups hiring out the Events Space to put on productions and activities, as well as a wide range of workshops and talks given to tenants: "[There's] lots of energy and creativity; the work comes naturally – plenty of cross-fertilization of ideas and opportunities" (ibid.: 2011). ·

Finally, while Coexist exemplifies the novel bottom-up innovations and community action initiatives generally becoming more visible in environmental politics (Seyfang & Smith, 2007), it would be a mistake to think that Coexist is simply about subaltern networks. The building's owners are substantial Bristol-based property developers who are heavily involved in social housing, green building, and other sustainability initiatives. Similarly, the high-profile Canteen restaurant, which occupies the ground floor of the building, was designed by a leading Bristol architect and cultural entrepreneur. Coexist has very explicitly tried to engage with established Bristol institutions and networks – including Bristol City Council, the University of the West of England, the University of Bristol, and the Bristol Environmental and Technology Services project – in an effort to gain formal recognition for their activities. Indeed, remember that this project arose out of Coexist's approaching the University of Bristol rather than vice versa, and the need for mutual benefits from this engagement has remained central to subsequent discussions.

There is also interest in the idea of an incubator for social enterprises that would replicate the high-tech model of business incubation by providing space, mentoring, and access to business development opportunities. The aim is to link technological innovation and community action in ways rarely evident at present and may allow a new generation of micro-scale "social entrepreneurs" to become business entrepreneurs in due course. Finally, there is also an ambition for "Coexist Collaborates," an advisory service that could help others – nationally and internationally – learn from the Coexist example. This step would see Coexist's bottom-up initiative both scale up and begin to travel, becoming part of the heterogeneous "policy mobilities" (Peck & Theodore, 2010) that characterize contemporary processes of political learning and experimentation.

Fostering Resilient Spaces and Subjects

While at first glance Coexist might appear as a paradigmatic, grass-roots, community-based initiative, it should be clear by now that the organization does not neatly fit this category. It does not receive direct state funding – indeed, this option is explicitly rejected by those involved – but neither is it run according to market principles. Indeed, like many contemporary organizations, Coexist has emerged above a reconfigured governmental and political terrain in which key actors regularly work across conventional sectoral boundaries. It may be better thought of as an example of the "flex organizations" (Wedel, 2004) that are becoming widespread after neoliberalism. These organizations assemble multiple resources and sources of power across what were once understood as discrete domains. While Wedel's analysis focuses on how flex organizations escape conventional forms of accountability, the broader point is that these organizations are able to represent themselves in multiple forms. Certainly the Coexist directors appear equally as comfortable with property developers, local authority members, senior university managers, and engineers as they are with the young artists, activists, and creative practitioners who occupy Hamilton House.

The ability to move relatively fluidly among different communities and across sectors is important for their success. In this regard the Coexist directors exemplify the skills and attributes of the actors known as "strategic brokers" (Larner & Craig, 2005), or "translators" and/or "transactors" (Freeman, 2009; Newman, 2012). They tended to come from non-traditional, often activist or alternative, backgrounds. They did not draw on the professional knowledges traditionally associated with community development (such as planning, social work, social policy). Rather, they emphasized their non-expert engagements underpinned by creativity and ingenuity and privileged bottom-up solutions as part of a more generalized move towards sustainability:

> We believe this [empowering individuals and communities] can be achieved by working together to harness the skills already present within communities. Coexist don't think this is a hard task. In fact we are seeing that it can be done. We are all already entirely capable of innovation, of clarity and to effortlessly coexist with ourselves and with each other. Each of us is a pioneer. Each of us has a unique contribution to the whole. And each moment is ours to make the most of. Here, coexisting naturally, we see that together we are capable of so much. (Coexist director, 2011)

This non-traditional approach crosses over to the operation of the building which is based on participatory and deliberative processes that invite self-management and self-organization. Initially there was a stipulation that all applicants for space in Hamilton House would agree to volunteer for Coexist for one day a week. This is no longer a condition of entry, in part because it proved unworkable, but the organization continues to privilege non-hierarchical leadership, transparency in business dealings, mutual support, and, above all, empowerment of all the individuals participating in the project to become involved and improve their own working and personal lives. There is a strong emphasis on the communal aspects of the building, including weekly tenant lunches, and even today word of mouth continues to be the means by which new tenants are identified. As they stress in their official documentation, "Taking an unbureaucratic, can-do approach allows authentic and relevant creativity to flourish" (Moreton, 2011: 3).

This approach and this ethos are not simply about building new forms of social connectivity. Coexist is attempting to stitch together diverse forms of activity and political engagements – such as those found in alternative therapies, creative practices, and sustainability initiatives – in a supportive space that also facilitates economic opportunities for its tenants. In short, Coexist is not an anti-capitalist initiative, despite the frequent evocation of the language and concepts of anarchism. They are very clear that their wider aim is to generate new economic opportunities, to create jobs and in doing so to help make the transition to sustainable communities and a low-carbon economy. But their aim is to do so in a way that challenges more established conceptions of economic development. As outlined in their vision statement by Jamie Pike, Coexist founding director, "We have grown organically and will continue to do so, open to respond to what is needed by the community within and without the building. Our direction and action has a single purpose – to support innovative solutions for all. This means personal, social and collective, within and without the space we facilitate, and for current and future generations" (Hamilton House, 2014).

Cultivating resilient subjects is understood to be central to this process. Being in Hamilton House not only gives all those involved a space within which to work on their own projects, it also explicitly fosters innovation, relationality, and skill sharing. The Coexist directors themselves (who change frequently and sometimes unexpectedly) occupy multiple roles and are engaged in diverse activities and interests. Nor are these activities and interests necessarily associated with the traditional community sector and/or grass-roots politics. In addition to being part of Coexist, the directors are part of wider social entrepreneurial networks locally and nationally. The Coexist directors are open about their own personal development and support the development of their staff, enabling them to launch their own projects (inside and outside Hamilton House), write their own job descriptions, and develop a variety of skills, such as management, interpersonal skills, event facilitation, event organizing, and budget management.

More generally, Hamilton House is full of people seeking alternatives to conventional careers, identifying possibilities that generate a living wage in a context where they neither can have nor desire steady jobs and that at the same time allow them to "live their dreams." The tenants describe an experience of "serendipitous encounters" between "crazy arty people all doing their own thing." Indeed one tenant stated that his experience of the building was "reasonably shambolic." He went on to explain that this was not a criticism, but rather the dynamic/creative space he expected, and it created opportunities for his development. In his formulation being resilient meant that he could imagine and embody flexibility and adaptability within unpredictable settings and processes. Many of the tenants also privileged reflexivity and the need to be attentive to bodies and personal relationships, reflecting the wider emphasis on self-help and self-esteem that has emerged out of social movements and critical feminist practices.

As will already be apparent, not only are precarious, unstable, insecure forms of working and living being actively embraced, but they are also being used to open up new possibilities for those involved. Focus group discussions uncovered numerous examples of tenants who have located and developed opportunities for others among the wider Hamilton House community. Advice and networks are explicitly shared, often in informal settings, in ways that have a lasting impact. This sharing can involve discussions about potential career opportunities, which are especially important for self-employed individuals. More tangible examples include commissioning paintings for other organizations, finding independent video editors within the building who can assist another film production company, or self-employed people sharing advice on taxes and business structures. In many of these new organizational and occupational strategies attributes such as communication, affect, opinion, attention, and taste were explicitly named and actively engaged to create relationships with "people like us."

This is not a temporary phenomenon. Rather, these are the enduring occupational strategies found in wider arts ecologies (Markusen, 2006) and the creative industries (Gill & Pratt, 2010; McRobbie, 2011; Molloy & Larner, 2010) and that are now becoming generalized and more widespread. In these strategies being part of the "scene," active networking, and presentation of self are key to social participation and subsequent economic success. As is also well established in these wider literatures, such strategies are unlikely to follow conventional patterns of economy, work, and career. Rather than being a temporary stage on the way to more permanent arrangements, they involve a long-term engagement with self-employment, contingent labour, and/or "portfolio careers" (Kong, 2011). As we have already seen, for some this process is both energizing and exciting; for others it is hazardous and chaotic.

In sum, Coexist and those who have joined them in Hamilton House understand themselves as actively producing alternative futures and serving as role models for new modes of living and working. In this setting the explicit performance of creativity (which can be environmental, economic, or cultural), and the patterns of labour associated with that performance, are understood not as neoliberal exploitation but rather as an expression of the "living differently" that a more sustainable future will require. While the use of creative practice and alternative lifestyles as drivers for personal and social change is far from new, after neoliberalism they have becomes the means to new ends. Indeed, it can be argued that the alternative living and working conditions that emerged out of social movements of a previous generation have become increasingly economically viable in a context where precarity is now normalized, giving rise to what McRobbie (2011) has called "radical social enterprise."

In Coexist the fostering of radical social enterprise is associated with the explicit cultivation of the more resilient subjects who will make up the sustainable communities of the future. These processes of subjectification privilege personal reflexivity and collective empowerment in areas such as environment, health and well-being, and creativity. They are explicitly oriented towards wider forms of alternative living and transformative change, rather than the individualized techniques of economic entrepreneurship and active citizenship that underpin advanced liberal strategies of governing. They are also deeply communal and explicitly embodied, in comparison with the biological and behavioural approaches that increasingly shape fields such as public policy making and education (Pykett, 2013).

Conclusion

In organizations like Coexist and spaces like Hamilton House the attributes and qualities understood to be required for resilience are being explicitly named, learned, and worked upon to foster particular kinds of subjects. To quote Gibson-Graham & Roelvink (2010: 342), they are subjects who are reading the possible futures that are "barely visible in the present order of things" and imagining "how to strengthen and move them along." In Coexist, claims about the need to develop resilience are underpinned by discourses of creativity, networking, and ingenuity that offer support to diverse people in their attempts to engage with wider ambitions for sustainable communities. They are not the prudential subjects of neoliberalism who calculate the costs and benefits of risks of acting in a particular way. Nor are they the de-socialized, privatized, and individualized subjects of *homo economicus*. Instead, in this context resilience thinking emphasizes adaptability and flexibility, explicitly involves engaging with groups and networks, and privileges relationality.

More generally, the Coexist project underlines the need to examine how heterogeneous aspects of political life are being taken up and reconfigured in governmental formations that attempt to move beyond neoliberalism. It offers an empirical case that lays bare the continuities and disjunctures found in political and governmental formations that are still "in the making." While it might be argued that this case is overly particular, close attention based on ethnographic approaches has helped identify how resilience is being understood and how the spaces and subjects are being constituted. How are terms like "resilience," with its implications for how we organize our lives, constitutive of spaces and subjects in other settings? Answering this question demands an analysis that moves beyond easy, generalized claims about neoliberalism and after neoliberalism. Just as the shift from Keynesian welfarism to neoliberalism was associated with multiple new forms of subjectification, so too are the new interventions that ask us all to "resist shocks in a more responsive fashion" (Leach, Scoones, & Stirling, 2010: 60). A move to resilience may indeed mark a shift in our understanding of the present, but this movement is not singular.

REFERENCES

Adger, N. (2000). Social and ecological resilience: Are they related? *Progress in Human Geography*, 24(3), 347–64. http://dx.doi.org/10.1191/030913200701540465
Berkes, F., & Ross, H. (2013). Community resilience: Toward an integrated approach. *Society & Natural Resources*, 26(1), 5–20. http://dx.doi.org/10.1080/08941920.2012.736605

Berndt, C. (2014). Behavioural economics, experimentalism and the market-ization of development. Mimeo. Available from author.

Bristow, G. (2010). Resilient regions: Replacing regional competitiveness. *Cambridge Journal of Regions, Economy and Society, 3*(1), 153–67. http://dx.doi.org/10.1093/cjres/rsp030

Bristow, G., & Healy, A. (2014). Regional resilience: An agency perspective. *Regional Studies, 48*(5), 923–35. http://dx.doi.org/10.1080/00343404.2013.854879

Brown, K. (2014). Global environmental change I: A social turn for resilience? *Progress in Human Geography, 38*(1), 107–17. http://dx.doi.org/10.1177/0309132513498837

Bulkeley, H., Broto, V.C., & Edwards, G. (2012). Bringing climate change to the city: Towards low carbon urbanism? *Local Environment: The International Journal of Justice and Sustainability, 17*(5), 545–51. http://dx.doi.org/10.1080/13549839.2012.681464

Çalişkan, K., & Callon, M. (2010). Economization, part 2: A research program for the study of markets. *Economy and Society, 39*(1), 1–32. http://dx.doi.org/10.1080/03085140903424519

Chandler, D. (2010). Review article. Risk and the biopolitics of global security. *Conflict Security and Development, 10*(2), 287–97. http://dx.doi.org/10.1080/14678801003666024

Chandler, D. (2012). Resilience and human security: The post-interventionist paradigm. *Security Dialogue, 43*(3), 213–29. http://dx.doi.org/10.1177/0967010612444151

Christopherson, S., Michie, J., & Tyler, P. (2010). Regional resilience: Theoretical and empirical perspectives. *Cambridge Journal of Regions, Economy and Society, 3*(1), 3–10. http://dx.doi.org/10.1093/cjres/rsq004

Coaffee, J. (2013). Rescaling and responsibilising the politics of urban resilience: from national security to local place-making. *Politics, 33*(4), 240–52. http://dx.doi.org/10.1111/1467-9256.12011

Cooper, M. (2010). Turbulence: Between financial and environmental crisis. *Theory, Culture and Society: Explorations in Critical Social Science 27*(2–3): 1–24.

Cretney, R., & Bond, S. (2014). Bouncing back to capitalism? Grassroots autonomous activism in shaping discourses of resilience and transformation following disaster. *Resilience, 2*(1), 18–31. http://dx.doi.org/10.1080/21693293.2013.872449

Demos (2009). *Resilient nation*. Available at www.emergencymanagement.org.uk/portals/45/documents/Resilient-Nation-Executive-Summary.pdf. Accessed 1 April 2015.

Duffield, M. (2011). Environmental terror: Uncertainty, resilience and the bunker. Working Paper 06–11. School of Sociology, Politics and International Studies, University of Bristol.

Eraydin, A. (2013). "Resilience thinking" for planning. In A. Eraydin and T. Taşan-Kok (Eds), *Resilience thinking in urban planning* (17–37). Dordrecht, NL: Springer. http://dx.doi.org/10.1007/978-94-007-5476-8_2

Eraydin, A. & Taşan-Kok, T. (Eds). (2013). *Resilience thinking in urban planning*. Dordrecht, NL: Springer. http://dx.doi.org/10.1007/978-94-007-5476-8

Ettlinger, N. (2011). Governmentality as epistemology. *Annals of the Association of American Geographers, 101*(3), 537–60. http://dx.doi.org/10.1080/00045608.2010.544962

Evans, J. (2011). Resilience, ecology and adaptation in the experimental city. *Transactions of the Institute of British Geographers, 36*(2), 223–37. http://dx.doi.org/10.1111/j.1475-5661.2010.00420.x

Folke, C. (2006). Resilience: The emergence of a perspective for social-ecological systems analyses. *Global Environmental Change, 16*(3), 253–67. http://dx.doi.org/10.1016/j.gloenvcha.2006.04.002

Freeman, R. (2009). What is "translation"? *Evidence & Policy, 5*(4), 429–47. http://dx.doi.org/10.1332/174426409X478770

Fung, A., & Wright, E.O. (2003). Deepening democracy. London: Verso.

Gibson-Graham, J.-K. (2011). A feminist project of belonging in the Anthropocene. *Gender, Place and Culture, 18*(1), 1–21. http://dx.doi.org/10.1080/0966369X.2011.535295

Gibson-Graham, J.-K., & Roelvink, G. (2010). An economic ethics for the Anthropocene. *Antipode, 41*(1), 320–346. http://dx.doi.org/10.1111/j.1467-8330.2009.00728.x

Gill, R., & Pratt, A. (2010). In the social factory? Immaterial labour, precariousness and cultural work. *Theory, Culture & Society, 25*(7–8), 1–30.

Hamilton House (2014). Hamilton House. Available at http://www.hamiltonhouse.org/about/. Accessed 27 May 2015.

Hoffman, L. (2010). Patriotic professionalism in urban China: Fostering talent. Philadelphia: Temple University Press.

Joseph, J. (2013). Resilience as embedded neoliberalism: A governmentality approach. *Resilience: International policies. Practices and Discourses, 1*(1), 38–52.

Kong, L. (2011). From precarious labor to precarious economy? Planning for precarity in Singapore's creative economy. *City, Culture and Society, 2*(2), 55–64. http://dx.doi.org/10.1016/j.ccs.2011.05.002

Larner, W. (2000). Neoliberalism: Policy, ideology, governmentality. *Studies in Political Economy, 63*, 5–26.

Larner, W. (2006). Brokering citizenship claims: Neoliberalism, Tino Rangatiratanga and multiculturalism in Aotearoa, New Zealand. In E. Tastsoglou, & A. Dobrowolsky (Eds), *Women, migration and citizenship: Making local, national and transnational connections* (131–48). Aldershot, ON: Ashgate.

Larner, W. (2011). C-change? Geographies of crisis. *Dialogues in Human Geography*, *1*(3), 319–35. http://dx.doi.org/10.1177/2043820611421552

Larner, W. (2012). New subjects. In J. Peck, T. Barnes, & E. Sheppard (Eds), *The new companion to economic geography* (358–71). London: Wiley-Blackwell.

Larner, W. (2014). The limits of post-politics: Rethinking radical social enterprise. In J. Wilson and E. Swyngedouw (Eds), *The post-political and its discontents: Spaces of depoliticization, spectres of radical politics* (189–207). Edinburgh: Edinburgh University Press. http://dx.doi.org/10.3366/edinburgh/9780748682973.003.0010

Larner, W., & Craig, D. (2005). "After neoliberalism"? Community activism and local partnerships in Aotearoa, New Zealand. *Antipode*, *37*(3), 402–24. http://dx.doi.org/10.1111/j.0066-4812.2005.00504.x

Larner, W., Le Heron, R., & Lewis, N. (2007). Co-constituting "after neoliberalism?" Political projects and globalising governmentalities in Aotearoa, New Zealand. In K. England & K. Ward (Eds), *Neoliberalization: States, networks, people* (223–47). Oxford: Blackwell. http://dx.doi.org/10.1002/9780470712801.ch9

Leach, M., Scoones, I., & Stirling, A. (2010). *Dynamic sustainabilities: Technology, environment, social justice*. Pathways to Sustainability. London, Washington, DC: Earthscan.

Lentzos, F., & Rose, N. (2009). Governing insecurity: Contingency planning, protection, resilience. *Economy and Society*, *38*(2), 230–54. http://dx.doi.org/10.1080/03085140902786611

Li, T.M. (2010). To make live or let die? Rural dispossession and the protection of surplus populations. *Antipode*, *41*(s1), 66–93. http://dx.doi.org/10.1111/j.1467-8330.2009.00717.x

Lorey, I. (2011). Virtuoso of freedom: On the implosion of political virtuosity. In G. Raunig, G. Ray, & U. Wiggenig (Eds), *Critique of creativity: Precarity, subjectivity and resistance in the "creative industries"* (79–90). London: MayFly Books.

Markusen, A. (2006). Urban development and the politics of a creative class: Evidence from a study of artists. *Environment & Planning A*, *38*(10), 1921–40. http://dx.doi.org/10.1068/a38179

Martin, R. (2012). Regional economic resilience, hysteresis and recessionary shocks. *Journal of Economic Geography*, *12*(1), 1–32. http://dx.doi.org/10.1093/jeg/lbr019

McRobbie, A. (2011). Rethinking creative economy as radical social enterprise. *Variant, 41*, 32–3.

Molloy, M., & Larner, W. (2010). Who needs cultural intermediaries indeed? Gendered networks in the designer fashion Industry. *Journal of Cultural Economics, 3*(3), 361–77. http://dx.doi.org/10.1080/17530350.2010.506322

Moreton, S. (2011). About Coexist: 2011 report. University of Bristol and Coexist. Mimeo. Available from author.

Newman, J. (2010). Towards a pedagogical state? Summoning the "empowered" citizen. *Citizenship Studies, 14*(6), 711–23. http://dx.doi.org/10.1080/13621025.2010.522359

Newman, J. (2012). *Working the spaces of power: Activism, neoliberalism and gendered labour.* London: Bloomsbury Press.

O'Malley, P. (2010). Resilient subjects: Uncertainty, warfare and liberalism. *Economy and Society, 39*(4), 488–509. http://dx.doi.org/10.1080/03085147.2010.510681

Office of the Deputy Prime Minister (UK) (2004). *Egan review: Skills for sustainable communities.* Available at http://ihbc.org.uk/recent_papers/docs/Egan%20Review%20Skills%20for%20sustainable%20Communities.pdf. Accessed 1 April 2015.

Ong, A. (1999). Flexible citizenship: The cultural logics of transnationality. Durham, NC: Duke University Press.

Peck, J., & Theodore, N. (2010). Mobilizing policy: Models, methods and mutations. *Geoforum, 41*(2), 169–74. http://dx.doi.org/10.1016/j.geoforum.2010.01.002

Pelling, M. (2010). *Adaptation to climate change: From resilience to transformation.* Oxford: Routledge.

Pykett, J. (2013). Neurocapitalism and the new neuros: Using neuroeconomics, behavioural economics and picoeconomics for public policy. *Journal of Economic Geography, 13*(5), 845–69. http://dx.doi.org/10.1093/jeg/lbs039

Pykett, J., Jones, R., Whitehead, M., Huxley, M., Strauss, K., Gill, N., … & Newman, J. (2011). Interventions in the political geography of "libertarian paternalism." *Political Geography, 30*(6), 301–10. http://dx.doi.org/10.1016/j.polgeo.2011.05.003

Raco, M., & Street, E. (2012). Resilience planning, economic change and the politics of post-recession development in London and Hong Kong. *Urban Studies (Edinburgh, Scotland), 49*(5), 1065–87. http://dx.doi.org/10.1177/0042098011415716

Rankin, K.N. (2001). Governing development: Neoliberalism, microcredit, and rational economic woman. *Economy and Society, 30*(1), 18–37. http://dx.doi.org/10.1080/03085140020019070

Reid, J. (2012). The disastrous and politically debased subject of resilience. Available at http://www.daghammarskjold.se/publication/end-development-security-nexus-rise-global-disaster-management/. Accessed 7 April 2012.

Rose, N. (1999). *Powers of freedom: Reframing political thought.* Cambridge: *Cambridge University Press.* http://dx.doi.org/10.1017/CBO9780511488856

Seyfang, G., & Smith, A. (2007). Grassroots innovations for sustainable development: Towards a new research and policy agenda. *Environmental Politics, 16*(4), 584–603. http://dx.doi.org/10.1080/09644010701419121

Shaw, K., & Maythorne, L. (2013). Managing for local resilience: Towards a strategic approach. *Public Policy and Administration, 28*(1), 43–65. http://dx.doi.org/10.1177/0952076711432578

Stumpp, E.-M. (2013). New in town? On resilience and "resilient cites." *Cities (London, England), 32*(1), 164–6. http://dx.doi.org/10.1016/j.cities.2013.01.003

Swyngedouw, E. (2009). The antimonies of the postpolitical city: In search of a democratic politics of environmental production. *International Journal of Urban and Regional Research, 33*(3), 601–20. http://dx.doi.org/10.1111/j.1468-2427.2009.00859.x

Wagenaar, H., & Wilkinson, C. (2015). Enacting resilience: A performative account of governing for urban resilience. *Urban Studies (Edinburgh, Scotland), 52*(7), 1265–84. http://dx.doi.org/10.1177/0042098013505655

Walker, J., & Cooper, M. (2011). Genealogies of resilience: From systems ecology to the political economy of crisis adaptation. *Security Dialogue, 42*(2), 143–60. http://dx.doi.org/10.1177/0967010611399616

Wedel, J. (2004). Blurring the state-private divide: Flex organizations and the decline of accountability. In M. Spoor (Ed.), *Globalization, poverty and conflict: A critical development reader* (217–30). Dordrecht, NL: Kluwer Academic.

Weichselgartner, J., & Kelman, I. (2014). Geographies of resilience: Challenges and opportunities of a descriptive concept. *Progress in Human Geography.* Advance online publication. http://dx.doi.org/10.1177/0309132513518834

Welsh, M. (2014). Resilience and responsibility: Governing uncertainty in a complex world. *Geographical Journal, 180*(1), 15–26. http://dx.doi.org/10.1111/geoj.12012

Zukin, S. (1982). *Loft living: Culture and capital in urban change.* New Brunswick, NJ: Rutgers University Press.

3 Governing through Community in the Oil Sands Zone

SARA DOROW

Jimmy and I stand looking at the river, downtown Fort McMurray behind us and some plumes of smoke visible in front of us.[1] "Somewhere about '96 or '97, my friend from one of the big construction firms calls me up and says, 'It's coming. All hell is going to break loose,'" says Jimmy. "And he was right." What Jimmy's friend saw coming was the explosive growth of the oil sands industry of northeast Alberta, Canada, built around the globe's third-largest proven oil reserve. For sixty-something-year-old Jimmy, who had lived through the earlier boom of the 1970s, this was a whole new ball game. New "market-based" tax and royalty policies introduced by the federal and provincial governments in the late 1990s – part of a broader set of neoliberal policies aimed at liberalizing trade and curtailing environmental regulation (Chastko, 2004) – had boosted oil sands production into the status of national "economic engine." By 2008, when Jimmy was interviewed, annual investment in the oil sands (a sticky mix of bitumen, sand, and water also known as the tar sands) had increased more than fivefold to nearly CAD$21 billion per year.[2]

It was in this decade that a neoliberal rationality asserting the synergistic compatibility and necessary partnering of economic and social pursuits took hold under the banner of "the Canadian Way" (Brodie, 2002). One key component of that new relationship was community. By the early 2000s, as Brodie remarks, "the idea of community as a partner in governance and the place where social needs are met [had] come to dominate the federal government's thinking" (391). A state no longer focused on "society" was to help diverse "communities" become strong and self-sufficient. Community had thus emerged as a new territory on and through which individuals would be governed, in what Rose (2001: 4) terms "ethopolitics": a set of governing strategies that act on the ethical dimension of human life, fostering the self-managed engagement of individuals in their collective destiny. Under the ethopolitical contract between official powers and subjects, "the former must provide the conditions of the good life, [and] the latter must deserve to inhabit it by building strong communities and exercising active responsible citizenship" (ibid.). In this way, community is a key node in governing the "complex composed of men and things" (Foucault, 1991: 93), that is, in realigning the relationships of individual subjects to both market and state (see also Bulley, 2013; Larner & Craig, 2005; Li, 2007; Staeheli, 2008).

When I first got off the bus in downtown Fort McMurray on a cold February day in 2007, my mind was on the complex of men (and women) and things (and places) in the resource extraction economy. What particular, context-specific configuration of ethopolitical practices and languages could be at work in a resource-dependent region that was materially and symbolically ascending as a bedrock industry in Canada? And how could an ethnographic study in Fort McMurray, the urban service area to the 64,000-square-kilometre municipality under which lies the Athabasca Oil Sands formation, bring to life the capillary and everyday phenomena of what I had begun to think of as a "neoliberal resource economy"?

Answers to these questions took many ethnographic field trips to Fort McMurray from my home base in Edmonton some 400 kilometres to the south, but they begin with two key points. First, if government is the "contact point" between technologies of self and technologies of domination (Foucault, Martin, Gutman, & Hutton, 1988; Lemke, 2002), where desires, habits, and aspirations are configured to align social and ethical life with economic criteria (Li, 2007; Ove, 2013; Vrasti, 2011), the particular political economic circumstances of resource extraction pose a unique set of governing problems. The concentration of productive and financial opportunities, the indenturing of the state and of citizenship to petro capital, the complex flows of labour, and the stresses of boom and bust that characterize the oil sands economy make for an overdetermined relationship between formal powers and self-determination (see also Sawyer, 2004; Shever, 2012). Under these particular conditions, the *homo economicus* (see also Li in this volume) of the oil sands needs governing strategies aimed less at increasing human capital[3] and more at taming, directing, and elevating its flow and quality.

Second, if community is "a territory between the authority of the state, the free and amoral exchange of the market, and the liberty of the autonomous, rights-bearing individual" (Rose, 2001: 6), the extra premium put on community in a context of close and fraught interdependence between private industry and everyday life can shed new light on Rose's analysis. In Fort McMurray what Rose (2001: 14) poses as two of four intersecting axes of government – "the emergence of community as an object of government" and "new specifications of political subjects" – become even more tightly bound as two sides of the same governing coin precisely because, in the shadow of oil, both the possibility of community and the freedom of the subject are in question. Reconstructing citizens as moral subjects of responsible community (ibid.: 1) thus depends on community being made the moral object of aspirational citizens of oil. Several years of ethnographic research in Fort McMurray have brought me to understand this as the imperative of "investing in community." In a context where subjectivization is otherwise so obviously beholden to the singular formal power of the oil industry, technologies that animate the moral will to *make community out of oil* foster an ethics of the free subject.

Investing in community is a constellation of strategies that entail both formal and informal techniques, and that enable contact and congruence between the "autonomous" working subject of Fort McMurray and the forms of political rule and economic exploitation (Lemke, 2002) that mark resource production.

Together, these strategies are part of the "topology" (Collier, 2011a) of governmental forms particular to spaces like the oil sands zone, where political rationalities of resource extraction – especially the direct governing power of private industry – are articulated to neoliberal notions of the free, entrepreneurial, ethical subject. By fostering aspirations to make community out of oil wealth here, where the oil is, these strategies do the work of aligning *citizen-for-community* and *citizen-for-market* (Larner & Craig, 2005; see also Venn, 2009) – or what Major & Winters (2013) in their study of Fort McMurray poignantly dub "community by necessity." As I argue below, the case of Fort McMurray extends Rose's (1996, 2001) original formulation by foregrounding the relevance of private enterprise, physical territory, and various forms of "inclusive exclusion" to the *how* of governing through community.[4]

Ethnographic Forays

The ethopolitics of converting human capital and oil wealth into community weds everyday practices and political economic norms by working on individuals' aspirations and subjectivities. Ethnographic work is well suited to apprehending these governmental strategies and to breaking down distinctions between macro- and micro-scales (Clough, 2010). By tracing links between formal programs and strategies and the messy, informal discourses and practices by which those programs are enacted and reimagined (Brady, 2014; Li, 2007; Vrasti, 2008), ethnographic inquiry mitigates the danger of any one case becoming yet another merely nuanced example of the same old neoliberalism (Vrasti, 2011). As a number of scholars have warned, the idea of neoliberalism as one inevitable thing must be countered through examination of the actual, messy constitution of a disparate array of neoliberal ideologies and practices (Larner, 2000; Sparke, 2006). The goal, then, is to apply an ethnographic sensibility to interpreting and representing the historical, political, and geographical relationships that articulate subjects with *specific* places, identities, and processes (Hart, 2004). A number of ethnographic studies of the neoliberal rationalities that are specific to resource zones have emerged in recent years, including Elana Shever's study of the realignment of communal and familial relations with the newly privatized oil industry in Argentina (see also Sawyer, 2004; Watts, 2004). Shever's project, like that of Tania Murray Li's (2007) *The Will to Improve*, helps to bridge a gap found in the governmentality literature between studies that focus on regulation (especially of southern peoples) and studies that focus on the constitution of ethical subjectivities (mostly in the north) (Ove, 2013: 317), but there remains very little ethnographic work that bridges this gap in northern resource-dependent communities (Young & Matthews, 2007).

My understanding of "investing in community" as a set of formal and informal ethopolitical strategies emerged from the ethnographic relationships among the researchers (myself and a team), our questions, the particular trails we followed, and the diverse array of people, events, and texts we encountered. Initial fieldwork began in 2007 at the height of the boom and continued through the downturn of late 2008 and into a new upswing by early 2010 (as is to be expected in resource economies; another downturn ensued in 2014). Most research trips to Fort McMurray occurred in 2008 and 2009 and lasted from a few days to several weeks. Our team started by conducting a set of sixty local "baseline" interviews and pursuing the unpredictable narratives that emerged from perambulatory participant observation. We then tackled a more strategic set of some thirty interviews with industry, municipal, development, and community leaders (any names used here are pseudonyms). It was largely from the latter interviews with people in positions of institutional authority and their discursive relationships to the broader "field" of everyday life in the region that I came to understand "investing in community" as an ethopolitical imperative specifically suited to the political rationalities of the resource production zone, where industry plays a key role in governing individual well-being and collective destiny.

Industry and Community in Fort McMurray

How the strategies of investing in community take shape, and how they align subject and community, are perhaps peculiar to Fort McMurray and the oil sands production zone, or at least to similar booming resource economies. The population of Fort McMurray doubled to 70,000 people in the early 2000s as "all hell broke loose" in oil sands development, making the region a national and international work destination for people in trades, industry professions, retail and hospitality, and social and public services. Some 60 to 70 per cent of work in the area is directly or indirectly in the oil industry, and there is high turnover and mobility. While many people call Fort McMurray home, and Cree, Dene, and Métis people have claimed it as home territory for generations, many others either fly in and fly out from homes elsewhere or arrive on the infamous "three-year plan" (a temporary move aimed at intensively exploiting work and money-making opportunities). Along with this growth and mobility, Fort McMurray and smaller rural communities throughout the region have experienced many of the precarious heights characteristic of booming resource-dependent communities: long working hours, high pay and cost of living, increased crime and drug use, and ever-increasing pressures on social and physical infrastructures (see Dorow & O'Shaughnessy, 2013).

These pressures place, and emplace, an intensified value on "quality of life" – availability of housing, voluntarism, leisure time, family time, and the possibility of a diverse, sustained social life – and thus, in turn, on the imperative to invest in community. While many of the official industry and community stakeholders we interviewed deployed a win-win rhetoric that posited a stable oil industry and a thriving community as mutually reinforcing ("capital costs are actually a function of the quality of life in the community," said one oil company executive), market and citizen always seemed on the verge of flying apart. The dominant presence of a large, single-resource industry undermined the neoliberal promise of folding market imperatives and quality of life into one another (Lemke, 2002; Vrasti, 2011) in two interrelated ways: first, by threatening to govern too much, and second, by offering an abundance of economic opportunity.

First, the resource extraction industry takes on a local role akin to the enabling or facilitating state (Rose, 2001: 6), in that it sets the conditions for local life not only through the provisioning of work opportunities and flows of capital, but also through flows of information and expertise, the provisioning of collective participatory and leisure opportunities, and the affective imbrication of individuals and collectives with the entrepreneurial state of oil (see also Hönke, 2012; Shever, 2012).[5] Thus, the oil sands industry – as much as if not more than formal state institutions – faces the risk of governing too much. Both community and industry leaders agreed that industry thus had a responsibility to enact but also actively delimit its intervention in the fostering of the local good life. One community service worker complained that "the industry idea of development doesn't transfer to community development," and another asserted that the oil sands industry's purpose "is to bring people here to work. Our motivation is to make a community." For their part, when asked about their role in community building, industry executives were apt to say, "be careful what you wish for," highlighting instead how they performed their own self-limit by indirectly acting through other governing bodies and organizations. These "facilitating" activities included, for example, supplying information about expansion plans so that municipal and community groups could also properly plan for social and infrastructural needs, lobbying the provincial government to direct royalty and tax revenues back to Fort McMurray and the surrounding communities and engaging in local philanthropy through third-party bodies such as the annual United Way campaign.

Second, however, the oil industry's performance of a self-limited, enabling role is persistently challenged by the singular cornucopia of economic opportunity it offers in the forms of investment capital, tax revenues, and work. During the five years of our study, the Canadian Association of Petroleum Producers and the federal government under Prime Minister Stephen Harper regularly lauded the diffusion of wealth in communities across the country made possible by Canada's "economic engine," including the entrepreneurial opportunity to move or at least commute to Fort McMurray to build and reap the promises of oil bounty.[6] As our interviewees saw it, an economic promise this charmed and this charged posed several specific challenges to governing the conversion of "opportunity for all" into local "responsibility and community" (Rose, 2001: 3, citing Tony Blair). There was the risk of squandering rather than multiplying the aspirational, life-giving possibilities of oil wealth in the very place it was produced. One city official mused that some day he was going to look back and say either "Man, I had a part in that!" or "Oh boy, did we ever screw that up." This risk upped the ante on techniques aimed at "building capacity," as discussed below. There was also the problem of the geo-territorial dispersion of work and life that came with high in- and out-migration and long-distance commuting ("those people pay taxes elsewhere and won't settle here with their families!" was a common gripe). As a result, government further ran up against the de-linking of production and consumption. Where Rose (2001) sees the fusion of culture and consumer activity as key to governing an ethopolitical collective, in Fort McMurray consumptive energies and identities were often tied to other places and/or deferred to the future (Dorow & Dogu, 2013). A minimal local consumer market ("the mall is too small!" complained the teens; "there are so few good restaurants!" complained the adults) went hand in hand with the resource economy's all-encompassing productive ethos and high labour mobility, requiring, respectively, the techniques of "consumptive attachment" and "giving back" discussed below.

To summarize, the booming oil industry in Fort McMurray is deemed a direct threat to the very quality of life promised by its abundance, putting an extra premium on community as the territory on and through which to make "economic enterprise coterminous with the energies of social life" (Vrasti, 2011). As Shever (2012) demonstrates, amid the remoteness, mobility, quick money and dependence that characterize oil extraction zones, community reactivates cultural values and sutures people to oil. In Fort McMurray strategies inciting investment in community are so vital because they do not merely react to the chaos and pressure of boomtown growth, but rather, proactively and even passionately take advantage of the professed opportunities, cultivating the "utopia" promised by the resource supply zone (Bridge, 2001). Economy and daily life become re-engaged via the governable practices and spaces where homo economicus, as both entrepreneur of herself and subject of domination, engages in the repeated, never-ending work of making community, including the very will to do so (cf. Li, 2007). Furthermore, these dynamics of individual, community and oil depend in part on the insider/outsider status of particular groups of people

(see also Shever, 2012; Staeheli, 2008) to convert the exclusionary implications (formal inequalities) of the resource economy into energies for
community. These processes work through spatial and affective affilia-
tions that are both racialized and gendered (Simon-Kumar, 2011; Vrasti,
2011).

Converting Oil Wealth into Community via the Investing Subject

> SARA: What do you think Fort McMurray will be like in ten years?
> INDUSTRY REPRESENTATIVE: Well hopefully, it's producing more oil from
> the oil sands but it's doing it in a way that's sustainable from a com-
> munity perspective – that there's more people who want to make it
> home and raise their families there.

As many official stakeholders in Fort McMurray see it, oil productivity is converted to sustainable community when individuals convert economic opportunity into the local good life. But this conversion
becomes a particular sort of problem in the oil sands zone, where governing community-as-territory is pulled inward by productive forces,
pulled outward by high worker turnover and mobility, and persistently nagged by the imperative to "catch up" to growth or at least to
mitigate its chaotic pressures. A truck driver who grew up in the area
described an atmosphere where "everybody's in their own little world,
just zooming around here trying not to crash into one another." A college instructor who had been in Fort McMurray for two years characterized the ambiance as a frustrating lack of commitment despite the
opportunities offered by oil: "There's a lot of people that come in and
feel like they're just here to get what they can out of it, big money, and
they say, 'Oh, I'll be here for two years.' Well that ends up sometimes
being fifteen years, but in those fifteen years they've never established
themselves, they've never given back to the community ... What I love
about Fort McMurray are the job opportunities. The opportunities for
your family ... You don't have to look far to find wonderful things here.
It's just a matter of making the effort."

Many of the formal programs directly employed by industry – that is,
providing information, lobbying, and philanthropic giving – can barely
address the problems of squandered opportunity, de-territorialized
relations of work and life, and diffuse cultural and consumptive identities that confront governing in Fort McMurray. The constellation of
strategies I call "investing in community" do this work, both formally

and informally setting the ethopolitical conditions for subjects to convert the opportunity *of* this place into opportunity *for* this place. In the sections that follow I explore three different but related strategies of investment that were revealed through our ethnographic fieldwork; in each case subjects that actively seek to make community out of oil are conditioned in part through the example of those who are positioned at the edges of belonging. The first of these strategies, which I call *investing in home*, conditions consumptive attachment, in part through the collective figure of the male mobile worker who is so fundamentally unattached. The second strategy, *investing by "giving back,"* conditions the conversion of oil-given opportunity, in part through the collective figure of the immigrant woman who desires to build and belong to community. And finally, *investing in capacities* conditions self-reliance for a future beyond oil, in part through the collective figure of the Aboriginal community that entrepreneurially seeks to remake its territorial relations through participation in the oil economy.

Conditioning Consumptive Attachment: Investing in Home

In a context of high mobility, turnover, and growth, investing in home becomes an important ethopolitical strategy for conditioning attachment. One important sign of making Fort McMurray home was the act of home buying itself. At the time of our research, the oil companies were under pressure to demonstrate their commitment to having direct and contracted employees move to the area rather than fly in and out on daily and weekly flights from other cities in Canada, such as Calgary or Kelowna or St John. The industry experimented with different approaches to deputizing their employees as good consumer and community citizens who would participate in the local housing market. An administrator at Syncrude, a major oil company, proudly asserted, "We have housing and retention and pay programs based on a local workforce ... So all job offers are: this position is a Fort McMurray based position." Other companies paid for several months' rent when employees first moved to Fort McMurray and/or offered moving bonuses to help cover down payments on a house – especially as the average cost of a house soared to over CAD$500,000. These techniques were intended to counter what one industry public relations office called the Fort McMurray lottery: "Selling a house [in Fort McMurray] for twice what they paid for it, moving their family to [a suburb of Edmonton or] some other nice community. You know, drive and fly back and forth, which is not good for quality of life in this community. It's a personal decision that they make, and which we can't change, but we can certainly offer incentives for them not to make that decision."

One company, CNRL (Canadian Natural Resources Ltd), even bought up 40 per cent of a new housing development in Fort McMurray to sell to its own employees. Called Eagle Ridge, this "livable community" development began in 2008 with restrictive rules for homeowners – for example, no renting and fixed percentage price increases for resales within five years – which the developer claimed were "designed to avoid speculation, flipping or hoarding in the tight market circumstances in Fort McMurray."[7]

Eagle Ridge represented one attempt to modify homo economicus from "only here to make money," and thus detached from place, to an investor of material and symbolic energies identified with "community life." The regional municipality and the provincial government also introduced programs that promised to condition people to invest in a home: affordable housing and cost-of-living allowances for public employees, incentives to buy in spite of a looming bust, and the creation of a public-private community development board to guide a large new housing development called Parsons Creek. A number of industry and municipal representatives saw this as the provincial government finally doing something to set the economic conditions – a delicate balance of government intervention and free-market forces – for ethopolitical investment in homes by Fort McMurray's working subjects.

However, these are fragile strategies that run up against the space-time realities of the resource extraction economy. A number of industry tradespeople told us that their work schedules (e.g., seven days on, seven days off) directly contradicted the push to buy a house locally and move their family to Fort McMurray. And many oil company employees with whom we spoke had made their housing bonuses an integral part of the "three-year plan": buy several housing units in town, rent them out, then sell them for a healthy profit when exiting Fort McMurray (provided the boom was still on). This pattern exacerbated local complaints that industry practices were directly contributing to the high cost of housing, and that neighbourhoods were being overtaken by renters or short-term buyers who filled driveways and streets with their trucks but refused to plant themselves in the community. One long-time resident pointed to the disinclination to literally plant in many Fort McMurray neighbourhoods: "People will put some flowers in a big pot out front, but they won't bother with planting trees." It was for these reasons that the moral enshrinement of investing in houses as an aspirational act of attachment was a crucial ethopolitical strategy.

Contracted mobile trades workers – the lone men and some women who lived in basement suites or in work camps near industry project sites – posed a particular threat to investing in homes. They were understood as disinterested in buying a house and moving their families to the area, and worse yet, as a *drain* on the community. Yet as essential productive labour in the industry that was the community's "bread and butter," mobile workers somehow had to be accommodated in techniques of investing in homes. Such accommodation was accomplished by deeming them contrastive outsiders to the community and thus as possible "converts" to it, that is, as an object both against and upon which the aspirational virtues of the investing subject of Fort McMurray could be exercised. We might see them as an inversion of Rose's (2001: 12–13) "excluded" – potentially "failed" individuals that become a collective target of ethically reconstructive programs of inclusion such as welfare-to-work – in that it was their very waged productivity that separated them from, and put a strain on, the community.

Some governing techniques were aimed at actively containing mobile workers. During the five years of our fieldwork, industry investment in additional and improved work camp facilities outside Fort McMurray were increasingly posited as good for the local community. By absorbing mobile work as an "operations issue," as one executive put it, industry funds expended on work camps and on the nearby airstrips that efficiently flew workers in and out of job sites could protect Fort McMurray and other local population centres from the onslaught of mobile workers who would otherwise live there, park three trucks deep in driveways, tie up traffic, and drive up the cost of housing with their living-out allowances. This same executive also pointed out that he and his peers would sometimes police the public behaviour of mobile workers by calling up contractors whose company trucks were seen weaving down the road: "There are things we can do as citizens, and I just happen to have perhaps a little more pull, that I can make a difference in the community." If properly contained, mobile workers could "free" territory for the subjects of Fort McMurray to properly invest in homes.

At the same time, other governing techniques made mobile workers an object of inclusion: by working to "engage" mobile workers, various non-profit groups and community stakeholders activated their own aspirations to convert oil wealth into the local good life. One long-time resident described the organization of a hockey game with camp dwellers, another regularly invited them to participate in music jams, and members of the local Leadership Wood Buffalo Program enthusiastically researched ways to reach out to the "shadow population" as part of the municipality's 2009 participatory branding campaign, called Big Spirit. What seemed to matter here was less the probability of attaching mobile trades workers to Fort McMurray and more the opportunity to aspire to helping them do so.

Conditioning the Opportunity to Belong: Investing by "Giving Back"

In Fort McMurray, the singular power ascribed to the oil sands industry to provide jobs and wealth – locally, but also, to some extent, provincially and nationally – had the effect of placing subjects in a deficit moral position laden with the responsibility to harness the general condition of opportunity. Their entrepreneurialism was to be evident in what they *did* with the opportunities provided by the oil sands economic engine. In Foucauldian terms, it was this context of opportunity that was to make an individual recognize her moral obligation.

Volunteering time and skill was a key ethopolitical mechanism (see King, 2003) through which humans were to convert abundant economic opportunity into community – an activity dubbed "giving back." One leader in Fort McMurray's non-profit sector described this as a double movement of selflessly giving individual time and attention to the very community that gave so much to the country:

> There's so much going on in this town, there's so many different busy schedules, busy families ... everybody just feels so all over the place. And then [people find] a way to stop and slow down and take a minute to realize what this community has done ... This town financial-wise has given so much to the country and there's so many different small non-profit organizations that provide programs and services that only benefit the community and there's so many people that only want to do something for somebody else. And to me, that's giving back because it's never for them.

Time was ascribed a particular premium in this moral economy, given long work hours and the sometimes mad pace of development. But if time was "taken away" by the very industry that provided such economic abundance, its duty as a governing institution was at least to help set the conditions for individuals to "give back." In this vein industry representatives I interviewed focused not on corporate voluntary or fundraising events, but rather on programs they had instituted to incite employees to be voluntary entrepreneurs of themselves. The two major oil companies, Syncrude and Suncor Energy, for example, awarded funds to non-profit organizations where individual employees volunteered on their own time. One of their officers stated that this approach routed company wealth into community well-being via the pride of individual employees: "We know it is important to our employees that they can put the Suncor logo on something or say [to an organization where they volunteer], 'Here's the Suncor cheque.' So that strategically figures into our sponsorships on the community investment side as much as anything else. It's more: what are our *employees* proud of?"

Another executive saw evidence of the industry's cultivation of investing subjects in the simple fact that its recruitment of so many skilled professionals *to* this place was ripe with opportunity *for* this place – if only those employees could be encouraged to invest their time and skill in the community. The trick, in this moral economy, was to inspire individual volitional investment apart from, yet in response to, the work opportunities provided by the oil industry as the prevailing governing body.

In the case of volunteering, it was not greedy mobile trades workers but determined foreign-born migrant workers that served as a collective inside/outside to the ethopolitics of "giving back." Here were subjects who, it seemed, eagerly sought to convert the abundance offered in the region (and by extension, Canada) into bettering both themselves and the community. In 2008 Volunteer Wood Buffalo (later called Wood Buffalo Volunteers) profiled several immigrant women volunteers on their website: "'I have no family here,' Gladys says, 'nor do I have a lot of money to give, but what I can contribute is my time ... volunteering is also a way to express my gratitude to a country that has given me a very warm welcome.'"[8] These stories, as one non-profit sector manager put it, served as a chastening ideal to the Fort McMurray subject: "If they are new and they are so actively involved in the community, then why aren't people who have been here for years as actively involved? And they want to give so much and they are learning so much from it. It's just who they are. They want to give back. No, they don't have the money to do it, but they just want to give back."

According to Rose (2001: 8), the ethopolitics of community depend on the collectivization of multiple, diverse identities. But in this case a particular collective of "diverse outsiders" valorizes community as an object of investment: this group migrates to the heart of oil and converts its productive opportunity into community through investing what that community lacks, namely, time. Their honourable example motivates insider subjects to make good by "giving back."

This aspirational relationship depends on a collective racialized and gendered imaginary of the outsider who desires to be inside and who gains community "belonging" by giving to it. It was especially apparent in the case of live-in caregivers in Fort McMurray, who became a special target group: white and professional women encouraged these mostly Filipina women to actively participate in the community through educational and voluntary opportunities. The Hub Family Resource Centre featured this testimonial from a female employer of a live-in caregiver, Angel, on its online forum: "Not only has the Hub been an asset to the development of my children [but] it has helped Angel network with other nannies too. Angel is now part of the Fort McMurray Community in her volunteer work and with her peers. This has enabled her to adapt to our Canadian culture and make her feel like she is at home ... Since my husband and I work full time we do not get a chance to sit in and participate [in Hub programming], so Angel is able to teach us what she learned during the sessions!"[9]

Foreign-born labouring women hold a contradictory inclusive/ exclusive status in the ethopolitical landscape that is only in limited ways similar to that held by mobile trades workers. Both groups have mobilized themselves to seek the opportunities in the region, and both hold the potential to be brought into the fold of community by the ethical Fort McMurray subject. However, the two groups then diverge into different sorts of racialized and gendered insiders/outsiders of the community. The mobile white men who are there "only to make money," not to invest in community – who embody a sort of excess of productivity, of economy unhinged and excluded from vital life – become an object of possible inclusion in the community because of the threat they present to it; they are a deficiency that must be compensated for by the individual Fort McMurrayite. The brown women who "give back" – who embody the desire to convert economic opportunity to the vitality of communal life – become idealized subjects of investment under the right conditions of government; they are the very possibility for community to which the Fort McMurray subject should and must be made to aspire.

Conditioning Self-Reliance: Investing in Individual and Collective Capacities

The two previous strategies of investment incited individuals to re-territorialize local community in the present, to turn productive oil wealth into consumptive and volitional acts of community building. A third strategy focused on building a future community "beyond oil." Imagining and making such a future was a point of both excitement and anxiety for Fort McMurray's industry, local government, and community stakeholders. The sheer size of the oil sands deposits allowed some magical thinking about how long the community could rely on oil wealth (one city planner revelled in the thought of a 300-year planning horizon) and/or heightened the sense of responsibility to make the most of that abundance. At the same time, however, uncertainty lingered about how long the abundance of wealth and work could last: some day the resource would run out, and in the meantime its viability in the marketplace was subject to boom and bust as well as the vagaries of pricing, competition, and technology.[10]

Governing here became a problem of turning the entrepreneurial energies of the present into the multiplicative, forward-looking possibilities of self-managed well-being, that is, investing in individual and collective capacities. Both industry and municipal strategies included, not surprisingly, a focus on "value added" partnerships and related forms of self-reflexive organization (Jessop, 2002; Larner & Craig, 2005). One pivotal example arose during our fieldwork when Suncor Energy partnered with the United Way to open a shared non-profit hub called the Redpoll Centre. Importantly, this initiative was seen as investing oil money in a community future weaned of its social (if not economic) reliance on oil money: "At some point we would like the community to have a foundation that doesn't just rely on industry for funding on a pretty much ad hoc annual basis but actually goes after the big dollars in terms of personal contributions," said a Suncor representative. For many municipal leaders the future of community also depended on investing oil money into a future beyond oil, but one characterized by local economic diversity. A city official framed it this way: "I'd like to know that if twenty-five years from now there was an alternative fuel or suddenly oil has no value anymore and these [oil sands] guys all close their doors and go home ... Can we have a sustainable community that's going to be able to grow because we have a university and all of our box stores and our tourism businesses and because there's so much [other] opportunity up here?"

A subtle battle for hearts and minds ensued between industry and the municipality over who was to set the conditions for future well-being, and how. Nowhere was this more apparent than in the strategic bid to govern investment in the capacities of regional hamlets populated by Aboriginal people (referred to in local parlance as "rural communities"). Given the encroachment of oil development on their traditional lands, and perhaps because they represented in both material and symbolic ways the existence of local community *before* oil, First Nations and Métis peoples became morally crucial to setting the conditions for community *beyond* oil. A municipal social planner remarked that, when it came to "rural sustainability planning," it was appropriate for the municipal government to partner directly with Aboriginal communities to help them imagine what kind of future they wanted apart from industry. "Maybe those community groups will go to industry to ask for certain things, right? But ... as a government, we really need to be standing strong." Diversifying options for Aboriginal communities was a particularly poignant facet of the municipality's capacity-building strategy – one that imagined that an economically diverse future beyond oil would arise, in part, from fostering the generative possibilities of a culturally diverse present.

Industry, meanwhile, poured resources into formal and informal relationships with Aboriginal leaders and their communities. Many major industry players published annual reports on what has come to be called "Aboriginal relations," and they regularly advertised how much they invested in Aboriginal communities. Nearly every industry executive I interviewed assigned a special place to "building capacity in [Aboriginal] communities around governance, around leadership, around youth, around health, and [around] education," as one of them related. This ethical commitment served to demonstrate that "Fort McMurray isn't just this little black hole somewhere where all we do is just go in and extract oil," noted another, but rather, that industry as a formal power was fostering appropriate and specific conditions for a prosperous communal and environmental future. The conditions, or capacities, for Aboriginal self-reliance were to be achieved through training, jobs, and the wealth that would follow – quintessential neoliberal strategies meant to proactively cultivate a transformative "market citizenship" (Altamirano-Jiménez, 2004) or what Ove (2013) conceptualizes as "developmentality." An official from Syncrude emphasized that a multi-million dollar partnership with the local college was the

company's proactive response to a request from Aboriginal stakeholders for more training opportunities. Industry representatives further
highlighted stories of self-driven Aboriginal employees hired and
trained by the industry who then later started businesses in their own
communities. This was described with some pride by one industry
executive as a generative cycle of "local" (a rhetorical stand-in for First
Nations and Métis people) capacity building: "We've lost an employee
in the process, but we've gained an Aboriginal business, which in turn
has created more employment opportunities for local people in the
region."

Collectively racialized as deep "insiders" in constant danger of being
excluded from the bounty of oil, Aboriginal communities become a
combination of object and subject for the ethopolitics of investing in
individual and collective capacities (see Friedel, 2008). But given the
importance of physical territory to the future of the oil industry, strategies that foster the conversion of oil sands wealth into the capacities
of Indigenous peoples must do so in a way that separates "Indigenous
culture and self-government from territory" (Altamirano-Jiménez,
2004: 350). A decade into the boom, Fort McKay (a hamlet of Dene,
Crée, and Métis people surrounded by bitumen mining sites, where
the river water has become undrinkable) had become something of a
poster child for capacity building and even an object lesson for other
Aboriginal groups. "Just look at what McKay is doing," oil executives
would say, pointing to the local businesses and industry partnerships
that were allowing the band office to replace dilapidated infrastructure
with brand-new homes and community buildings. Further touted was
Fort McKay's ability to pay forward economic opportunity by offering construction contracts to companies from Aboriginal communities
outside the oil sands zone. Moral inspiration for life beyond oil was
in this way provided by subjects who, rather than de-territorialized
by oil developmentality, were building capacity for the future through
entrepreneurial investment in the physical re-territorializing of their
home communities.

In Conclusion: Community as Territory in Fort McMurray

Ethopolitical strategies such as "investing in community" become
important to neoliberal government because political and economic
stakes in citizenship are not enough: the problems of human society
are "increasingly made intelligible as ethical problems" (Rose, 2001: 4).

How they become intelligible as ethical, and thus how they re-territorialize community, is a question of context and contingency.

Resource economies like the oil sands extend and to some degree revise the characterizations of the ethopolitical rationality of community laid out by Rose in his reflections on the politics of the "third way." First, the dominance of this singular (if multi-headed) industry – and one that must be here, where the resource is – lends it a formal position vis-à-vis community that threatens to undermine "governing at a distance" (see also Hönke, 2012); the deep imbrication of community with the oil economy both unsettles and overdetermines it as a territory where individuals "freely" choose and self-manage, suggesting that institutional economic arrangements are more important to the shape of ethopolitical rationalities than Rose's work implies. Second, some of the characteristics of the resource extraction zone, such as concentrated economic production, rapid growth, high labour mobility, and a promising but deeply uncertain future, destabilize the coherent "localness" of the very communities that are to benefit from oil's life-giving opportunity; here, governing through community requires more *geo*-territorial attention to the spatialized detachment, disruption and dispersal of identificatory ties than Rose seems to allow (see also Watts, 2004). Third, the excesses of capital and opportunity that mark a booming mega-program of resource extraction only selectively rely on the sort of strategies aimed at including individuals in economically productive life that are emphasized by Rose (2001); the formal and informal techniques aim at "investing in community" instead, inciting subjects to make the most of the opportunity so abundantly given. Finally, and relatedly, community as the conditioning grounds for the Fort McMurray subject depends on collective others that are in various states of unbelonging relative to the promised bounty of Canada's "economic engine"; unlike the failed (poor, indigent) economic subjects of "third-way" style neoliberalism, whose reattachment to community is meant to conjure access to work (ibid., 12), these collective figures are targets for and/or exemplars of conjuring community from the abundance of work. Together, these insights suggest that community-as-territory plays an important role in "adjusting" the contact point between technologies of domination and technologies of self.

By orienting me "to a group of sites and a set of problems that I simply could not have stumbled upon otherwise" (Collier, 2011b: 29), ethnographic exploration in an oil extraction zone revealed specifically, and in rich detail, why and how community is a crucial territory upon which, and a strategic object *through* which, free subjects, private industry, and "actual" local territory are flexibly articulated with each other. In Fort McMurray, concern for the dangerous effects of the centripetal and centrifugal forces of industry on local community exist alongside concern for the equally dangerous possibility of squandering its bounty. By conditioning subjects that choose to convert the opportunity *of* this place into opportunity *for* this place, strategies aimed at investing in community help to produce a "novel pattern of correlation between choice mechanisms and social welfare" (Collier, 2011a) in the oil sands economy. Ethnographic forays into the specificities of these circumstances help to conceptualize community as territory for governing relations between the ethical subject of neoliberalism and the political rationalities of resource enterprise zones.

ACKNOWLEDGMENTS

The research project "Social landscapes of neoliberal growth: The case of Fort McMurray" was funded by the Social Sciences and Humanities Research Council of Canada (ID 410-2008-432) and by a Killam Cornerstone Grant from the University of Alberta.

NOTES

1 In May 2016, as this book was going into production, a large wildfire destroyed parts of the city of Ford McMurray and caused some of the major oil sands companies to shut down operations temporarily. How the aftermath of that fire might foreground or reshape discourses and techniques of community remains to be seen.

2 Found at www.energy.alberta.ca/OilSands/791.asp (accessed 20 December 2014).

3 In the *Birth of Biopolitics*, Foucault (2008: 231) suggests that governing strategies are aimed at increasing human capital, which for neoliberalism is understood to stave off the falling rate of profit.

4 I am inspired here by Agamben's (1998) *Homo Sacer*, which highlights "inclusive exclusion" as a governing political rationality.

5 I call this a "corporate town" model in contrast to the "company town" model of direct provision of housing and other basic infrastructure that

has been classically described in works such as *The Company Town in the American West* (Allen, 1966).

6 Foucault (2008) argues that *homo economicus* mobilizes himself to build and/or develop his human capital (230–3).

7 PowerPoint presentation entitled "Centron-Residential-Presentation-to-WBBA-Feb-9-06" (accessed 19 March 2009).

8 Found at http://woodbuffalovolunteers.ca/.

9 Found at http://www.thehubfrc.ca/.

10 While the forty-year reign of the Progressive Conservative Party in Alberta suggested there was no cause for worry about the vagaries of formal provincial politics, the May 2015 victory of the social-democratic New Democratic Party (NDP) introduced a new governing variable for the oil sands zone.

REFERENCES

Agamben, G. (1998). *Homo sacer: Sovereign power and bare life*. Translated by Daniel Heller-Roazen. Stanford: Stanford University Press.

Allen, J. B. (1966). *The company town in the American west*. Norman: University of Oklahoma Press.

Altamirano-Jiménez, I. (2004). North American first peoples: Slipping up into market citizenship? *Citizenship Studies, 8*(4), 349–65. http://dx.doi.org/10.1080/1362102052000316963

Brady, M. (2014). Ethnographies of neoliberal governmentalities: From the neoliberal apparatus to neoliberalism and governmental assemblages. *Foucault Studies, 18*, 11–33.

Bridge, G. (2001). Resource triumphalism: Post-industrial narratives of primary commodity production. *Environment & Planning A, 33*(12), 2149–73. http://dx.doi.org/10.1068/a33190

Brodie, J. (2002). Citizenship and solidarity: Reflections on the Canadian way. *Citizenship Studies, 6*(4), 377–94. http://dx.doi.org/10.1080/1362102022000041231

Bulley, D. (2013). Producing and governing community (through) resilience. *Politics, 33*(4), 265–75. http://dx.doi.org/10.1111/1467-9256.12025

Chastko, P. (2004). *Developing Alberta's oil sands: From Karl Clark to Kyoto*. Calgary: University of Calgary Press.

Clough, P.T. (2010). The case of sociology: Governmentality and methodology. *Critical Inquiry, 36*(4), 627–41. http://dx.doi.org/10.1086/655203

Collier, S.J. (2011a) Foucault, assemblages, and topology. Interview with Simon Dawes. Available at http://www.theoryculturesociety.org/

interview-with-stephen-j-collier-on-foucault-assemblages-and-topology/.
Accessed 15 May 2015.

Collier, S.J. (2011b). *Post-Soviet social: Neoliberalism, social modernity, biopolitics.* Princeton: Princeton University Press. http://dx.doi.org/10.1515/9781400840427.

Dorow, S., & Dogu, G. (2013). The spatial distribution of hope in and beyond Fort McMurray. In T. Davidson, O. Park, & R. Shields (Eds), *Ecologies of affect: Placing nostalgia, desire and hope* (271–92). Waterloo, ON: Wilfrid Laurier University Press.

Dorow, S., & O'Shaughnessy, S. (2013). Fort McMurray, Wood Buffalo, and the oil/tar sands: Revisiting the sociology of "community." *Canadian Journal of Sociology, 38*(2), 121–38.

Foucault, M. (2008). *The birth of biopolitics: Lectures at the Collège de France, 1978–1979.* London, New York: Palgrave Macmillan. http://dx.doi.org/10.1057/9780230594180.

Foucault, M. (1991). Governmentality. In G. Burchell, C. Gordon, & P. Miller (Eds), *The Foucault effect: Studies in governmentality.* Chicago: University of Chicago Press.

Foucault, M., Martin, L.H., Gutman, H., & Hutton, P.H. (1988) *Technologies of the self: A seminar with Michel Foucault.* Amherst: University of Massachusetts Press.

Friedel, T.L. (2008). (Not so) crude text and images: Staging Native in "big oil" advertising. *Visual Studies, 23*(3), 238–54. http://dx.doi.org/10.1080/14725860802489908

Hart, G. (2004). Geography and development: Critical ethnographies. *Progress in Human Geography, 28*(1), 91–100. http://dx.doi.org/10.1191/0309132504ph472pr

Hönke, J. (2012). Multinationals and security governance in the community: Participation, discipline and indirect rule. *Journal of Intervention and Statebuilding, 6*(1), 57–73. http://dx.doi.org/10.1080/17502977.2012.655569

Jessop, B. (2002). Liberalism, neoliberalism, and urban governance: A state-theoretical perspective. *Antipode, 34*(3), 452–72. http://dx.doi.org/10.1111/1467-8330.00250

King, S.J. (2003). Doing good by running well: Breast cancer, the race for the cure, and new technologies of ethical citizenship. In J.Z. Bratich, J. Packer, & C. McCarthy (Eds), *Foucault, cultural studies, and governmentality* (295–316). New York: SUNY Press.

Larner, W. (2000). Neo-liberalism: Policy, ideology, governmentality. *Studies in Political Economy, 63*, 5–25.

Larner, W., & Craig, D. (2005). After neoliberalism? Community activism and local partnerships in Aotearoa New Zealand. *Antipode, 37*(3), 402–24. http://dx.doi.org/10.1111/j.0066-4812.2005.00504.x

Lemke, T. (2002). Foucault, governmentality, and critique. *Rethinking Marxism, 14*(3), 49–64. http://dx.doi.org/10.1080/089356902101242288

Li, T.M. (2007). *The will to improve: Governmentality, development, and the practice of politics.* Durham, NC: Duke University Press. http://dx.doi.org/10.1215/9780822389781.

Major, C., & Winters, T. (2013). Community by necessity: Security, insecurity, and the flattening of class in Fort McMurray. *Canadian Journal of Sociology, 38*(2), 141–66.

Ove, P. (2013). Governmentality and the analytics of development. *Perspectives on Global Development and Technology, 12*(1-2), 310–31. http://dx.doi.org/10.1163/15691497-12341257

Rose, N. (1996). The death of the social? Re-figuring the territory of government. *Economy and Society, 25*(3), 327–56. http://dx.doi.org/10.1080/03085149600000018

Rose, N. (2001). Community, citizenship, and the Third Way. In D. Meredyth & J. Minson (Eds), *Citizenship and cultural policy* (1–17). London: SAGE. http://dx.doi.org/10.4135/9781446218990.n1.

Sawyer, S. (2004). *Crude chronicles: Indigenous politics, multinational oil, and neoliberalism in Ecuador.* Durham, NC: Duke University Press. http://dx.doi.org/10.1215/9780822385752.

Shever, E. (2012). *Resources for reform: Oil and neoliberalism in Argentina.* Stanford: Stanford University Press.

Simon-Kumar, R. (2011). The analytics of "gendering" the post-neoliberal state. *Social Politics, 18*(3), 441–68. http://dx.doi.org/10.1093/sp/jxr018

Sparke, M. (2006). Political geography: Political geographies of globalization (2) – governance. *Progress in Human Geography, 30*(3), 357–72. http://dx.doi.org/10.1191/0309132506ph606pr

Staeheli, L.A. (2008). Citizenship and the problem of community. *Political Geography, 27*(1), 5–21. http://dx.doi.org/10.1016/j.polgeo.2007.09.002

Venn, C. (2009). Neoliberal political economy, biopolitics and colonialism: A transcolonial genealogy of inequality. *Theory, Culture & Society, 26*(6), 206–33. http://dx.doi.org/10.1177/0263276409352194

Vrasti, W. (2008). The strange case of ethnography and international relations. *Millennium, 37*(2), 279–301. http://dx.doi.org/10.1177/0305829808097641

Vrasti, W. (2008). "Caring" capitalism and the duplicity of critique. *Theory & Event, 14*(4). http://dx.doi.org/10.1353/tae.2011.0041

Watts, M.J. (2004). Antinomies of community: Some thoughts on geography, resources and empire. *Transactions of the Institute of British Geographers*, *29*(2), 195–216. http://dx.doi.org/10.1111/j.0020-2754.2004.00125.x

Young, N., & Matthews, R. (2007). Resource economies and neoliberal experimentation: The reform of industry and community in rural British Columbia. *Area*, *39*(2), 176–85. http://dx.doi.org/10.1111/j.1475-4762.2007.00739.x

4 Fixing Non-market Subjects: Governing Land and Population in the Global South

TANIA MURRAY LI

If government is the attempt to direct conduct and optimize relations between "men and things," it must be grounded in concepts of society and human nature and supported by expert knowledge about the particular domains of conduct that need to be adjusted. This kind of expert knowledge can be generated by various techniques (field research, surveys, polls, focus groups) generally aligned with ethnography, understood here not as a method (i.e., long-term fieldwork), but as the graph of ethnos: the scientific study of "peoples," "cultures," or populations. Particular understandings of human nature figure in the liberal and neoliberal rationalities of government that Foucault analysed in his later work. They also figure in governmental assemblages that combine these broad rationalities with more specific diagnoses and prescriptions to improve human conduct in a given socio-spatial milieu.

This chapter explores deployments of expert knowledge about specific populations in attempts to govern relations between rural populations and land in the global south. Since colonial times a peculiar and recurrent feature of these assemblages has been the attempt to fix particular types of land and particular types of people to a non-market niche. Surprisingly, the fix persists in neoliberal assemblages that engage everyone as *homo economicus* (see Dorow in this volume). The articulation of liberal and neoliberal rationalities helps to explain how we have arrived at homo economicus minus the market – the curious, collective subject that figures in contemporary land regimes and conservation initiatives.

I begin with a brief review of Foucault's analysis of liberal and neoliberal rationalities and the understanding of human nature embedded within them. Then I outline my concept of governmental assemblage and proceed to explore how liberal elements in colonial assemblages have been grafted onto neoliberal ones that attempt to govern through the particularities of communities and their capacity to make good choices. Note that my source of data for this chapter is not field based, and I do not examine the practices through which the assemblages I examine are pulled together or what happens to them when they hit the ground, the kind of "ethnography of governing" I have pursued in other work (Li, 2007a, 2007b). My main focus is on "ethnography in governing": the work done by expert knowledge about populations in assemblages that attempt to direct human conduct for improving ends.

Liberal and Neoliberal Rationalities and the Work of Assemblage

A liberal rationality of government, Foucault argues, is concerned with the economical management of society, understood as a natural system with its own mechanisms, in which intervention always has to be scrutinized from the perspective of the risk of governing too much (2008: 19). Graham Burchell summarizes this as "governing in accordance with the grain of things ... to the end of securing the conditions for an optimal, but natural and self-regulating function" (1991: 127).The task of government in liberal mode is to enframe social processes in mechanisms of security so they can take their natural course, adjusted only slightly. The assumption, in short, is that society pre-exists, and "it is the natural, self-producing existence of this society that the state has to secure" (140). Knowledge of "society" therefore becomes important: it is both the object of liberal government, and sets its limits.

As many scholars have shown, knowing "society" in the eighteenth and nineteenth centuries was, in significant part, a matter of "making it up" by devising categories, collecting statistics in relation to those categories, and measuring and comparing around a norm (Hacking, 1986; Rose, 1999). It also required expert knowledge of particular populations, especially those understood to deviate from the norm – paupers, criminals, people suffering from various diseases, and so on. The purpose of this knowledge was to support the liberal mode of governing "in accordance with the grain of things" because the grain was not uniform. To govern particular populations appropriately meant knowing their character and evaluating their capacities. Expert knowledge became the basis for governing some groups in a liberal manner, that is, through their intrinsic capacity for autonomous, self-regulating conduct.[1] Groups found to not possess the necessary capacity – women, children, the sick, paupers, other races, colonial subjects – would be governed differently, in idioms of trusteeship, wardship, benevolent protection, paternalism, and often in an authoritarian or despotic manner (Foucault, 2007: 8–9; Hindess, 2001; Mehta, 1997; Procacci, 1991; Valverde, 1996). If we were to conduct a head count, Barry Hindess suggests, the "liberal government of unfreedom" was far more common, even in the metropoles, than the ideology of liberalism suggests (2001: 101). Designated groups, and the particular assemblages that were pulled together to govern them, formed what Uday Mehta(1997) describes as the constitutive exclusions of liberal rule.

The knowledge required to govern distinct populations in accordance with their nature was extensive. In India and other colonial situations, distinct populations were made known through scientific studies that became the basis for formulating appropriate modes of governing, and were embedded in law, producing variations on "apartheid" themes.[2] These studies were also needed to justify liberal strategies of exclusion and to respond to persistent, liberal critiques that intervention in the lives of colonial subjects could be excessive. Uday Mehta argues that James Mill's purpose in writing his massive, seven-volume *History of British India* was to demonstrate India's "exceeding difference" in depth and in detail – expert knowledge that liberal government demanded as grounds to justify India's permanent subjection (1997: 78, 60).

Neoliberal rationalities maintain the liberal emphasis on governing "in accordance with the grain of things," but the knowledge requirements of this mode of government are different. The axiom or truth at the centre of the neoliberal rationality of government is a view of human nature drawn from game theory. It is explicitly universal: humans are homo economicus. They decide upon a course of action according to a calculation of costs and benefits. Hence, neoliberal governing consists in setting conditions and devising incentives so that prudent, calculating individuals and communities choosing "freely" and pursuing their own interests will contribute to the general interest as well (Read, 2009: 27–9; Rose, 1999). These interests are not necessarily connected to making a profit through market transactions: actors rationally pursue diverse goals (public service, finding a spouse, tending a garden). Governing is a matter of "getting the incentives right" so that some conduct is encouraged and enabled, while other conduct becomes more difficult. Hence, to govern in a neoliberal style it is necessary to discover the contextual factors that rational actors incorporate into their decisions and, where necessary, adjust them.[3] Expert knowledge about contextual factors is furnished through quantitative techniques to measure behaviour (surveys and statistics) and through qualitative techniques (interviews, observations, focus groups) that explore how actors understand the costs and benefits of the choices before them. Neoliberal governing also goes along with intensified procedures for audit and accountability and constant, detailed, on-the-ground checking to ensure that perverse incentives have not been accidently introduced.

Foucault's late lectures (2008) explored what it means to govern according to a particular rationality or ethos. He distinguished between liberal and neoliberal rationalities and made further distinctions within the neoliberal ethos as it took shape differently in Germany, France, and the United States. In probing these variants, he began the work of situating governmental rationalities in the specific socio-spatial, historical milieu in which they take on their particular form. A "governmental assemblage," as I use the term, emerges in such a milieu. It is the field of knowledge, practices, and devices from which particular programs of intervention are derived. It is assembled under a dominant governmental ethos or rationality – a characteristic way of understanding the work of government – although more than one rationality may be present, because rationalities do not operate in a vacuum (Hindess, 1997: 266; Lemke, 2002: 55, 57). Rationalities are inscribed in practices or systems of practice that take shape in relation to problems to be solved, an accumulation of laws formulated in different eras, habits of thinking, inscription devices, material elements (trees, soil, water, labour, etc.), forms of knowledge, social relations, compromises, and critical responses to previous assemblages and their effects. Understanding how assemblages are pulled together and made to cohere requires attending to both rationalities (what makes it rational to think in this way, to proceed in this manner) and the work of assemblage, which involves managing fractures, dealing with incoherencies and forging alignments. To explore the specificity of governmental assemblages, the rationalities that animate them, and the struggle to make heterogeneous elements cohere, I now turn to an investigation of attempts to govern relations between land and people in the global south.

Liberal Government and the Differentiated Subjects of Colonial Rule

Colonial rulers in the nineteenth century were concerned with defining the proper role of markets in governing relations between land

and people. Hence, land relations became a key arena in which ideas about how to govern distinct subsets of the population were worked out. Should colonial subjects be understood as competent and potentially prosperous market subjects, capable of taking care of their own improvement and paying taxes? Or should they be protected from market pressures they were culturally unsuited to manage well? How could the protection or improvement of colonial subjects be made compatible with other colonial imperatives, namely, the allocation of land for plantations and settlement, essential sources of revenue and profit? Colonial authorities devised varied solutions to this problem. A common approach was to divide the population into types, each of which could be governed according to its nature.[4] In schematic form, colonial dividing practices and the resulting assemblages were something like the following:[5]

1) One part of the population – usually urban and educated – was deemed fit to be governed in a liberal manner. They could hold land individually, sometimes with formal, legal titles.
2) Another part of the population – call them farmers – could be exposed to market discipline, which was understood as an educative device, a vehicle for instilling habits of frugality, industry, and prudent calculation. Their individual land rights were recognized, and they often paid taxes based on land.
3) Another part – call them peasants – were incompetent or feckless; hence, they needed to be protected from the full force of market discipline by interventions to make land inalienable, control interest rates, and forbid usury. These measures were necessary to prevent them from dispossessing themselves and creating problems of destitution and disorder. With proper guidance some members of this group might "graduate" to join group (2) above.
4) A fourth part of the population – call them tribes – were understood to be so different that they could not be educated or improved, and they definitely should not be exposed to market risk. Instead, they should be governed in terms of their difference: fixed in their alterity, fixed in place on ancestral/customary land, and fixed in the sense of repaired, where inappropriate changes would be reversed or abated to restore them to their authentic selves.

Colonial scholars and administrators focused much of their attention on the fourth category, which provided grist for understanding

alterity in binary terms. If the colonizers recognized individual property, tribal others must hold communal/collective rights not equivalent to "property." If the colonizers were driven by profit motives and
sought private gain, tribal others must be inclined to share, and so on.
The differences aligned on this binary axis could be read as disabili-
ties or as virtues, since "they" have what "we" have lost. Difference
provided the grounds for fixing tribal others in place, preserving their
authentic otherness or sometimes clearing them out of the way, so they
could be replaced by more efficient land users.

Specifics varied.[6] To take just one example: in colonial United States
after the Dawes Act (1887) Indians could obtain individual title to their
land. They were no longer to be treated as a distinct type en masse.
Their degree of difference and the decision about who could or could
not obtain title to their land were made on an individual, case-by-case
basis. People who the Indian agents determined were capable of managing their own market destiny were set free to pursue it, while those
who agents deemed to be incompetent continued to have their land
managed by trustees. This was clearly a case of graduation: a racialized
disability could be overcome with enough paternalistic tutelage, expo-
sure to white ways through the introduction of European settlers onto
reserve land, and the encouragement of intermarriage. A consequence
of individual title – under conditions of structural/racial violence – was
that a great number of Indians lost their land (Biolsi, 2005).

In some parts of Asia, namely, India, Vietnam, and the Philippines,
colonial authorities divided the population on a spatialized basis in terms
of elevation. People living in highland areas became defined as tribes
and were subject to distinct legal regimes that ran under the umbrella
of custom. Their land was declared communal and inalienable. These
are the people who later came to be classified as "Indigenous People" in
international conventions. Those living in lowland areas were treated
as peasants and were entitled to hold (and perhaps lose) their land as
individuals. But the use of law as a tactic to fix identity and govern
conduct did not always work. People forbidden from selling land did
so anyway, to cover debt, but did so "illegally," hence at a disadvanta-
geous price. Land markets did not disappear; they were driven underground. Some people designated for the tribal slot actually desired or
had long held individual rights, and they resented paternalistic restrictions that did not repair their authentic state, but imposed a new and
unwelcome one. These people-land governing assemblages, in short,
were fractious and fragile. Both the people governing and the people

whose conduct was to be governed debated, disrupted, or unravelled them, and both sometimes acclaimed, supported, or demanded them when they were consonant with their own practices and desires.

In the Netherlands East Indies (later Indonesia), the Dutch colonial regime did not resolve the land dilemma by dividing the population into peasants and tribes. Instead, officials and legal scholars continued to debate the proper way to govern land relations until the end of Dutch rule. The "stumbling block," as critics describe it, was the racial axis of the apartheid legal regime that divided the colony's population into just three types: European, Foreign Oriental, and Native (Fasseur, 1994). The 1870 land law rendered all native landholding communal and inalienable, a move legislators understood to be in keeping with authentic native practice. The law's communal presumption was later amplified and confirmed by the prominent legal scholar Cornelis van Vollenhoven (1874–1933), who presided over the collection of huge volumes of scientific data about native law and custom.

Van Vollenhoven's research methods were not ethnographic in the modern sense, as he spent little time in Indonesia and was not exposed to awkward and dissonant encounters that might have challenged his views. He relied on materials collected by his students and associates, sifting through them to identify common threads. Indeed, he made a virtue of abstraction, arguing that only trained foreign experts could "discover" native customs, because native informants were merely practitioners, incapable of synthesis (Burns, 1989: 9, 97; Sonius, 1981). When drawn into the colonial land regime, expert knowledge about native communalism served to informalize or deny individual land tenure, which was widespread (Burns, 2004). Communal rights were prominent in the law, but in practice they were not well defined, nor were they secured through the necessary inscription devices (maps, lists, boundary markers). The failure to fix land or people left a legacy of confusion and acute vulnerability around land rights that continues today.[7]

Liberal Elements in Contemporary Assemblages

In the past few decades, liberal strategies of exclusion and paternalist protection that took shape in the colonial period have been appropriated and made into demands by "Indigenous" people for secure tenure over communal, inalienable land. For Indigenous activists and their allies this kind of ethnic-spatial fix serves as a bastion against dispossession by large-scale state and corporate projects: dams, mines, plantations, and the like. Building on their own traditions, the colonial legacy, and globally circulating narratives, rural people claiming the Indigenous slot anchor their racial/cultural difference in the radical alterity mode: love of Mother Earth, commitment to ancestral terrain, community, conservation, and sustainable livelihood practices. It is a strategy that carries its own dilemmas. Indigenous people who fail by the standards of radical alterity-as-virtue risk being treated as illegitimate, discarded by their allies and placed outside the distinct legal regimes designed to protect and govern them (Conklin & Graham, 1995; Li, 2000, 2001). A focus on the Philippines enables me to explore the Indigenous fix, and to situate it in relation to two other contemporary assemblages that attempt to govern relations between people and land.

The Philippines Indigenous Peoples Rights Act (IPRA) of 1997 provides an example of how popular demands framed in terms of alterity can be grafted onto a governmental assemblage, where they set limits on what people can do with their land (Bruelmann, 2012; Bryant, 2000). The act was the product of decades of struggle by Indigenous groups, activist lawyers, sympathetic politicians, and social movements. The rights granted under the act are significant achievements, but they are – necessarily – distinct from the rights that other rural Filipinos enjoy. IPRA land titles are communal and inalienable, and they carry the burden of "sustainable" management. The act treats Indigenous Filipinos as non-market subjects. Not only is their land to be kept off the land market, but their livelihoods should be limited to those that inflict little damage on the environment, a limitation not imposed on individual landholders or corporations. So governing Indigenous people differently, in accordance with their (assumed) culture and "natural" interest – that is, in a liberal manner – is bolstered by law, a coercive means to make sure that they do, in fact, do as they ought.

The contemporary assemblage that emerged to govern migrants who have moved into upland and forested areas of the Philippines is distinct. These migrants are not considered culturally "other," and they do not qualify as Indigenous under IPRA. Their distinction is the physical condition of the territory they occupy – sloping land – and its ecological and political status, since it is claimed by the forest department. Migrants insist on farming this land even though the authorities are convinced it should be forested to prevent erosion, mudslides, and other kinds of damage. The problem is that migrants behave as entrepreneurial market subjects, which is fine in the lowlands, but is not acceptable to the forest department because they are in the wrong place (Gauld, 2000; Hall, Hirsch, & Li, 2011).The resolution is a legal vehicle called the Community Forest Lease, which regularizes the de facto presence of these migrants in the state-claimed forest, but attempts to limit and direct it. To qualify for the lease, migrants are obliged to organize their affairs as rule-bound "communities" that must achieve specified standards of sustainability in their land and forest use (see Dorow in this volume). The attempt, in effect, is to make them into non- or limited-market subjects, rather like the Indigenous people who are the imagined subjects of IPRA. But the liberal argument that this approach simply restores them to their natural state cannot be sustained: scientific studies show that migrants on land frontiers tend to be entrepreneurial and seek to improve their livelihoods and, where possible, make a profit (Nelson, Cramb, Menz, & Mamicpic, 1998; Yap, 1998). Holding this assemblage together requires managing the fractures and contradictions, authoritarian rules and duress: the migrants accept the deal because it is better than being evicted.

The third, contemporary Philippines assemblage I will briefly examine is land reform. The land reform law grants individual title to land reform beneficiaries, but restricts it: the recipients are barred from selling their land. The rationale for this fix is incapacity. Decades of Philippines experience with land reform, scientifically confirmed, has shown that beneficiaries of land reform soon return to old habits of feckless behaviour (gambling, debt), or fall victim to unscrupulous moneylenders, leading to a downward spiral of land mortgage and eventual sale. Hence, they need to be encouraged to participate in markets as entrepreneurial farmers, but protected from market risk. Similar restrictions are applied to land reform beneficiaries in many contexts, where farmers must demonstrate suitably reformed conduct before they can "graduate" and be trusted to enter into markets for land (Hall et al., 2011; Hetherington, 2009).

These three contemporary assemblages for governing relations between people and land in the Philippines have distinctive features, as they were pulled together from elements available in this particular milieu: liberal concepts of difference, dividing practices and laws of colonial provenance, activist programs to defend peoples' access to land, transnational Indigenous rights discourse, features of the landscape (hills, forests, boundaries), Cold War experiences with land reform as a means to manage dangerous classes, an active Maoist insurgency, expert discourses, inscription devices (forest maps, satellite images,

farm surveys), laws and habits of legal thinking, and many other elements. To properly study how these elements are assembled and these
alignments are forged would require a more thorough analysis than I
can provide here, one that relies at least in part on field research and the
use of ethnographic methods.[8]

For the purpose of the present argument, what I want to stress is
the continuity of a liberal ethos of governing "in accordance with the
grain of things," as it works its way into various configurations. I also
want to stress the surprising persistence of non-market fixes, which
are not unique to the Philippines or simply a carry-over from the past.
They figure prominently in recent land laws and regulations across the
southeast Asian region and beyond. Alongside land-titling programs
for selected subjects, especially in cities and in the lowlands, we find
that particular populations (sometimes named Indigenous), and farmers who occupy specific types of places (highlands, state-claimed forest lands), are being protected from market processes deemed to be
unsuitable for them (Hall et al., 2011). This finding runs counter to
market-centred definitions of neoliberalism, which stress the extension of markets to all people, places, and things (Castree, 2010: 1728). It
directs our attention to the heterogeneous character of assemblages, in
which neoliberal rationalities do not operate alone.

Homo Economicus Minus the Market

Liberal rationalities for governing land and population in keeping with
their (putatively) distinct capacities persist, and are incorporated into
contemporary assemblages where they articulate more or less awk-
wardly with neoliberal rationalities centred on the figure of homo economicus. Governing through this rationale means setting conditions
and calibrating incentives so that individuals and communities who
weigh the costs and benefits of alternatives will make the right choice.
The concept of homo economicus is universal. It encompasses all actors,
pursuing any and all ends, not only narrowly economic ones. It does
not divide populations according to their differential capacity to act in
a rational, calculative manner, but it does recognize that populations
(often described as "communities") have different values, desires, and
calculations. The role of expert knowledge about specific populations in
assemblages formed through this rationale is to explore the ends such
populations pursue, the calculations they adopt, and the most effective
way of using incentives to achieve governmental or "improving" ends.

In terms of land relations, neoliberal assemblages put the emphasis on choice. They reduce paternalistic protections to reward efficiency and entrepreneurship. They have largely dismantled the category "peasant," the third in the schema I outlined earlier, arguing that the people so classified should join with other farmers who sink or swim "freely" in the tide of market competition. Indigenous people are also governed through choice, but differently. Building on the colonial legacy of governing them as collectivities, which often is enshrined in law, neoliberal governing treats Indigenous people as collective subjects or "communities" with a neoliberal twist: they are communities capable of exercising their "free, prior, informed, consent" (commonly known by initials and phrased verbally as FPIC). Indeed, Indigenous people may be addressed only as collective subjects: group members who insist on exercising rights as individuals effectively become non-Indigenous, as they fall outside the scope of laws and practices designed to engage and protect them. Community and homo economicus are brought even more closely into alignment in contemporary assemblages that combine an insistence on collective decision making with the incorporation of Indigenous communities as legal persons empowered to enter into commercial contracts (see Dorow in this volume).[9]

An understanding of Indigenous people as competent, decision-making, rights-bearing, collective subjects has not been unilaterally imposed. It is the product of decades of advocacy and popular demand, which culminated in the move to make FPIC a central feature of the 2007 UN Declaration on the Rights of Indigenous Peoples.[10] From now on Indigenous people – like all other people – are to be treated as autonomous and made responsible for their own fate. More specifically, like other people, they are empowered to make choices according to a rational calculation in contexts in which the conditions have been set to encourage some outcomes and discourage others. To make good choices they need good information, clear "rules of the game," and the opportunity to reflect on the costs and benefits of different courses of action. They may also need some fixing or repair through programs to train them in democratic procedures so that their choices will reflect an authentic, properly collective decision, not one that has been arbitrarily imposed by rogue members or outsiders seeking to manipulate their "choices" for private gain (Borras & Franco, 2010). The difficulty of distinguishing between setting conditions and manipulating outcomes is an obvious tension in these assemblages (Hirtz, 2003; Li, 2007b). Rose (1999) highlights the paradox, observing that community is assumed to pre-exist, yet it always needs to be fixed or perfected (see Dorow in this volume). Lemke extends this insight to neoliberal rationality in general, which he describes as "a political project that endeavours to create a social reality it suggests already exists" (2001: 203).[11]

The idea that communities/tribes/Indigenous people have distinctive characteristics, habits, and cultures is an element inherited from and formed within liberal assemblages and embedded in many legal regimes for land management. It remains available to be grafted onto neoliberal assemblages that draw on a different assumption: that tribes/Indigenous people, despite their differences, or in and through their authentic difference, are, in fact, capable of autonomous, rational, responsible choice. Concepts of difference are sustained in these assemblages because the nature of the "improvement" sought is distinct. Land management assemblages involving Indigenous people hinge on the assumption that they have desires, habits, and beliefs that favour collectivity, equity, environmental sustainability, and conservation. It is their difference (their indigeneity) as well as their sameness (their capacity for rational choice) that make FPIC necessary. Duly trained, fixed to be true to themselves, furnished with the appropriate information and given the right incentives, they can be counted on to make the right choices. This means they will reject practices that damage land, pollute water, or destroy forests. They will not privatize their collective land or plant it with lucrative cash crops, take on debt or sell land to the highest bidder. In terms of the schematic categories I outlined earlier, they will not try to become farmers, nor will they become peasants – people who aspire to be market subjects but fail because they make poor choices that cause trouble for themselves and others.

Conservation programs linked to the threat of global warming amply illustrate the neoliberal ethos running through assemblages for governing, still distinct but now empowered, capable, decision-making communities. For example, Forest Trends and Rights and Resources International aims to help Indigenous people secure their rights over forest land, so they can sell eco-system services such as biodiversity protection and carbon sequestration. They recognize the need and desire of Indigenous people to make a decent living from their land. But because the principal objective of these programs is conservation, they are clear about one thing: Indigenous people must not become entrepreneurial farmers with individual title to their land. If they acquire such a title, they will only lose it (Ellsworth & White, 2004). Instead, they are encouraged to engage in "community based agroforestry," a practice that both arborealizes and collectivizes their identities, live-lihoods and presumed preferences (Li, 2010; Walker, 2004). As far as possible, proponents of these programs want the character of "forest" people, places, and practices to be fixed permanently in law. The con-servation imperative, which is understood in global terms, means that "forest people" cannot be permitted to graduate.[12]

Interventions to fix Indigenous people to the land and limit them to specific land uses are currently being intensified in the context of billions of dollars in potential funding from the UN system, donors, and corporations for a program called REDD+: Reducing Emissions from Deforestation and Degradation, which is part of the effort to com-bat climate change. REDD+ is thoroughly neoliberal in its governing ethos. As one proponent put it, REDD+ is a matter of correcting for market failures and putting a price on ecosystem services, so that that "forest owners and users can simply sell forest carbon credits and less cattle, coffee, cocoa or charcoal" (Angelsen, 2010). From this perspec-tive, all that is needed to make REDD+ work is "the three 'I's ... Infor-mation, Incentives and Institutions" (ibid.). But these are not generic "I's." The extensive documentation about REDD+ highlights the need for detailed, scientific study to ensure that information, incentives, and institutions are tailored to the particular character of the local/forest/ Indigenous communities where they will be applied.

The demands of REDD+ are peculiar: market incentives (cash com-pensation) or collective benefits (schools, facilities) are to be given to forest communities and Indigenous people so that they will not culti-vate or indeed do anything at all with their land except conserve and restore forest. The hope (or article of faith) carried over from liberal thinking is that forest conservation is an activity for which Indigenous people and "forest communities" already are culturally disposed. This might mean they do not need to be paid so much or maybe not paid at all, since forest conservation is their own authentic goal, one they already factor into their calculations. But there is enough doubt among both proponents and critics of the REDD+ program that the commit-ment to FPIC is a guarded one.[13] The Indonesian REDD+ Task Force, for example, states that "Consultation is based on complete, balanced, hon-est, unbiased and easily understood information concerning the alter-natives and choices existing for the public within the implementation of REDD+ activities, along with the consequences of each alternative choice …This information is meant to create leeway for broad consen-sus, with all parties having access to existing opportunities" (2012: 28).

Note that "the public" (unspecified) will be given a choice, though this is a choice governed by the demand for "broad consensus," a notion often attributed to Indonesian culture and tradition, though in practice applied coercively to disallow dissent. The choices are among REDD+ alternatives, not outside them. It is not clear that forest villagers, whose participation is essential for the consensus vision of REDD+, actually have the option to make the "wrong" choice: to cut forests and to reject REDD+ programs and REDD+ funds in favour of different goals or different productive arrangements.

Doubts about the quality of "choice" and the risk of coercion are recognized in a report on REDD+ commissioned by the UN's Food and Agriculture Organization (FAO). "In order for FPIC to have real validity, communities must have a legally binding option not to consent. This binding framework will probably not emerge anywhere, which is why the World Bank uses the term 'consultation' rather than 'consent'" (Anderson, 2011: 39). The Indonesian program seems to address the risk of coercion by promising social safeguards to protect "vulnerable groups, including Indigenous peoples and local communities living in and around forests, whose livelihoods depend on forest resources"(Indonesian REDD+ Task Force, 2012: 29), but it is a form of protection that fixes them, once again, in their distinct identities and their arboreal niche. The program does not have room for the possibility that "vulnerable" groups might reject REDD+ in favour of more lucrative land uses, such as planting cash crops. REDD+ will not work if they graduate.

We see in this assemblage the attempt to enrol parties said to be "naturally" aligned with the program goals, extending to them the choice to which they are entitled, but attempting to channel it. Only "forest people" and Indigenous people who have been collectivized, arborealized, fixed in place, and fixed in their difference could be encouraged, empowered, and potentially coerced to make a non-market choice, one that commits them to making less money than they could with other "options." They are subjects of choice, but the assumption (or fragile hope) is that the good they will "naturally" choose to maximize is not profit: they are homo economicus, but their values and calculations are distinct. Nevertheless, proponents and critics are not naive, and they have serious doubts about whether the "3 I's" – information, incentives, and institutions – can be put in place.[14] Their doubts are often based on field research or, more generally, on deep immersion in forest bureaucracies characterized by corruption, collusion, rent-seeking, lawlessness, coercion, and a notable lack of even basic information, such as current forest or plantation maps. They also draw on scientific studies of forest communities, which show them to respond rather quickly and fully to changing incentives.[15] Indeed, to take homo economicus seriously as a universal is a sobering prospect: it is all too easy to get the incentives wrong, creating outcomes quite different from the ones desired.

Conclusion

The governmental assemblages I have examined in this chapter are guided by rationalities with distinct understandings of society, human nature, and the proper way of intervening in them. These rationalities and the forms of knowledge they generate are consequential: it matters whether rural people are understood as vulnerable children who need to be protected, as potential entrepreneurs waiting to be activated by incentives and the provision of micro-credit, or as a nature-loving tribes who would never sacrifice Mother Earth and future generations for short-term gain. As I have stressed, expert knowledge about the character of (putatively) distinct peoples, places, cultures, habits, and beliefs was intrinsic to liberal governmental assemblages, which aspired to govern in accordance with "the grain of things." It plays a different role in neoliberal assemblages, because homo economicus is assumed to be universal. Hence, the differences between actual populations are variations on rational-choice themes. Nevertheless, to ensure accountability, transparency, freedom, and choice – all the elements that homo economicus needs to operate efficiently – scientific research is still needed. Its task is to identify distortions and correct them, to devise suitable incentives, and to monitor carefully to make sure that consent is properly "free," and that the results conform to the plan.[16]

My focus in this chapter has been on the role of expert knowledge about populations in the formation of governmental assemblages, or "ethnography in government." I have not provided an ethnography of government, although I have done this work elsewhere through detailed, field-based studies of the practices through which governmental assemblages are pulled together and made to cohere. These practices included formulating and defending authoritative knowledge, managing contradictions, enrolling actors, building alignments, rendering political problems technical and calculable, and others. The scope for ethnographic methods in this field of inquiry is extensive: they can be used to study how knowledge about distinct communities is produced, evaluated, and assimilated or set aside; how concepts of difference are inscribed and made usable; how dissent and insurrectionary demands are managed, appropriated, or disallowed; how limits are drawn to what can or cannot be accommodated within an assemblage like FPIC; what is recognized as coercive – not free – and how it is acted upon; how tribes are resituated as vulnerable subjects requiring protection; and so on. Ethnographic techniques are also well suited for tracking the effects of such assemblages under particular conditions and the practices and subjectivities they produce and enable, which are often unexpected.

Actually existing governmental assemblages are shot through with contradictions that may cause the elements to fracture and realign. But they are not radically contingent: it is not the case that anything goes. The idea of governing Indigenous people through their free, prior, and informed consent would have been unthinkable in colonial contexts in the nineteenth century, where evolutionary and racialized thinking disallowed the idea that tribes were capable of making rational choices and taking responsibility for their own lives. Similarly, the degree of paternalism exercised then would not be compatible with contemporary notions of rights, freedom, and choice. As Barry Hindess observes, "neoliberalism is a liberal response to the achievements of the liberal mode of government," as subjects previously deemed deficient were brought into the domain of autonomy (1993: 311). This does not make it grounds for celebration, as is amply confirmed by the articulation of liberal and neoliberal rationales and recourse to authoritarian practices in the assemblages I have examined. Researching particular governmental assemblages and the rationalities that animate them offers a crucial vantage on the history of our present.

ACKNOWLEDGMENT

This chapter was first published in a 2014 issue of *Foucault Studies*, *18*, 11–33.

NOTES

1 The ambiguity in liberal thought between treating autonomous persons as naturally present and setting conditions to produce such persons is discussed in Hindess (1993).
2 Colonial research on particular populations did not necessarily involve field-based research methods. Its distinctive feature was attention

to cultural specificity, backed by the authority of science. Colonial ethnography is discussed in Dirks (2001), Moore (2005), Pels & Skalemink (1999), and Thomas (1994).

3 Governmental assemblages focused on setting incentives are discussed in Li (2007b).
4 These dividing practices and their liberal rationale are strikingly described in Hindess (2001).
5 I describe this dilemma and dividing practices in diverse colonial situations and provide references to support this schematic summary in Li (2010).
6 See, among others, Nally (2008), Berry (1993), Corbridge (1988), and Mamdani (1996).
7 See Down to Earth (2012), Warren (2005), Fitzpatrick (2007), Lev (1985), and Li (2007b).
8 For an exemplary ethnography of practices of assemblage see Bruelmann (2012).
9 Contra Read (2009: 36). On the compatibility of advanced liberal rule and "government through community," in which communities are both autonomized and responsibilized see Rose (1999).
10 The incorporation of popular demands into governmental assemblages is discussed in Larner (2003: 511).
11 See also Rose (1999: 177) and Li (2007b).
12 Compare the situation of Aboriginal populations in Australia, some of whom have succeeded in consolidating legal rights to their ancestral land and now have real choices: sign up with mining corporations or fall in with conservation agendas (Altman, 2013).
13 See the mix of hope, despair, and determination in the contributions to Angelsen (2009).
14 Angelsen (2010) and McGregor (2010). For further discussion of REDD+ see www.forestpeoples.org.
15 See McCarthy (2000), Peluso (1992), Levang (1997), and Feintrenie, Chong, & Levang (2010).
16 I examined the extensive use of ethnographic research techniques by officials in the World Bank planning initiatives in "community driven development" in Li (2007b).

REFERENCES

Altman, J. (2013). Land rights and development in Australia: Caring for, benefiting from, governing the indigenous estate. In L. Ford & T. Rowse

(Eds), *Between indigenous and settler governance* (121–34). New York: Routledge.

Anderson, K.E. (2011). *Communal tenure and the governance of common property resources in Asia: Lessons of experiences in selected countries.* Available at http://www.fao.org/3/a-am658e.pdf. Accessed 1 October 2015.

Angelsen, A. (2009). *Realising Redd+: National strategies and policy options.* Bogor, ID: CIFOR.

Angelsen, A. (2010). The 3 Redd 'I's. *Journal of Forest Economics, 16*(4), 253–6. http://dx.doi.org/10.1016/j.jfe.2010.10.001

Berry, S. (1993). *No condition is permanent: The social dynamics of agrarian change in sub-Saharan Africa.* Madison: University of Wisconsin Press.

Biolsi, T. (2005). Imagined geographies: Sovereignty, indigenous space, and American Indian struggle. *American Ethnologist, 32*(2), 239–59. http://dx.doi.org/10.1525/ae.2005.32.2.239

Borras, S.M.J., Jr, & Franco, J. (2010). Contemporary discourses and contestations around pro-poor land policies and land governance. *Journal of Agrarian Change, 10*(1), 1–32. http://dx.doi.org/10.1111/j.1471-0366.2009.00243.x

Bruelmann, I.W. (2012). *Ancestral domain: Land titling, and the conjuncture of government, rights and territory in central Mindanao.* Zurich: University of Zurich.

Bryant, R.L. (2000). Politicized moral geographies: Debating biodiversity conservation and ancestral domain in the Philippines. *Political Geography, 19*(6), 673–705. http://dx.doi.org/10.1016/S0962-6298(00)00024-X

Burchell, G. (1991). Peculiar interests: Civil society and governing "the system of natural liberty." In G. Burchell, C. Gordon, & P. Miller (Eds), *The Foucault effect: Studies in governmentality* (119–50). Chicago: University of Chicago Press. http://dx.doi.org/10.7208/chicago/9780226028811.001.0001.

Burns, P. (1989). The myth of Adat. *Journal of Legal Pluralism, 21*(28), 1–127. http://dx.doi.org/10.1080/07329113.1989.10756409

Burns, P. (2004). *The Leiden Legacy.* Leiden, NL: KITLV Press.

Castree, N. (2010). Neoliberalism and the biophysical environment 1: What 'neoliberalism' is, and what difference nature makes to it. *Geography Compass, 4*(12), 1725–33. http://dx.doi.org/10.1111/j.1749-8198.2010.00405.x

Conklin, B., & Graham, L. (1995). The shifting middle ground: Amazonian Indians and eco-politics. *American Anthropologist, 97*(4), 695–710. http://dx.doi.org/10.1525/aa.1995.97.4.02a00120

Corbridge, S. (1988). The ideology of tribal economy and society: Politics in Jharkhand, 1950–1980. *Modern Asian Studies, 22*(1), 1–42. http://dx.doi.org/10.1017/S0026749X00009392

Dirks, N. (2001). *Castes of mind: Colonialism and the making of modern India*. Princeton: Princeton University Press.

Down to Earth (2012). The struggle for land. *DTE Special Edition Newsletter*, December 2012. http://www.downtoearth-indonesia.org/story/dte-newsletter-93-94-full-edition-download

Ellsworth, L., & White, A. (2004). *Deeper roots: Strengthening community tenure security and community livelihoods*. New York: Ford Foundation/Forest Trends.

Fasseur, C. (1994). Cornerstone and stumbling block: Racial classification and the late colonial state in Indonesia. In R. Cribb (Ed.), *The late colonial state in Indonesia: Political and economic foundations of the Netherlands Indies, 1880–1942* (31–56). Leiden, NL: KITLV Press.

Feintrenie, L., Chong, W.K., & Levang, P. (2010). Why do farmers prefer oil palm? Lessons learnt from Bungo District, Indonesia. *Small-scale Forestry*, 9(3), 379–96. http://dx.doi.org/10.1007/s11842-010-9122-2

Fitzpatrick, D. (2007). Land, custom and the state in post-Suharto Indonesia: A foreign lawyer's perspective. In J.S. Davidson & D. Henley (Eds), *The revival of tradition in Indonesian politics: The deployment of Adat from colonialism to indigenism* (130–48). London: Routledge.

Foucault, M. (2007). *Security, territory, population: Lectures at the Collège de France, 1977–1978*. New York: Picador. http://dx.doi.org/10.1057/9780230245075.

Foucault, M. (2008). *The birth of biopolitics: Lectures at the Collège de France, 1978–79*. Basingstoke, UK: Palgrave Macmillan. http://dx.doi.org/10.1057/9780230594180.

Gauld, R. (2000). Maintaining centralized control in community-based forestry: Policy construction in the Philippines. *Development and Change*, 31(1), 229–54. http://dx.doi.org/10.1111/1467-7660.00153

Hacking, I. (1986). Making up people. In T.C. Heller, M. Sosna, & D.E. Wellberry (Eds), *Reconstructing individualism* (222–36). Stanford: Stanford University Press.

Hall, D., Hirsch, P., & Li, T.M. (2011). *Powers of exclusion: Land dilemmas in southeast Asia*. Honolulu: Univeristy of Hawaii Press.

Hetherington, K. (2009). Privatizing the private in rural Paraguay: Precarious lots and the materiality of rights. *American Ethnologist*, 36(2), 224–41. http://dx.doi.org/10.1111/j.1548-1425.2009.01132.x

Hindess, B. (1993). Liberalism, socialism and democracy: Variations on a governmental theme. *Economy and Society*, 22(3), 300–13. http://dx.doi.org/10.1080/03085149300000020

Hindess, B. (1997). Politics and govermentality. *Economy and Society*, *26*(2), 257–72. http://dx.doi.org/10.1080/03085149700000014

Hindess, B. (2001). The liberal government of unfreedom. *Alternatives*, *26*(2), 93–111. http://dx.doi.org/10.1177/030437540102600201

Hirtz, F. (2003). It takes modern means to be traditional: On recognizing indigenous cultural communities in the Philippines. *Development and Change*, *34*(5), 887–914. http://dx.doi.org/10.1111/j.1467-7660.2003.00333.x

Indonesian REDD+ Task Force (2012). *REDD+ national strategy*, Jakarta: Indonesian REDD+ Task Force.

Larner, W. (2003). Guest editorial: Neoliberalism? *Environment and Planning. D, Society & Space*, *21*(5), 509–12. http://dx.doi.org/10.1068/d2105ed

Lemke, T. (2001). "The birth of biopolitics": Michael Foucault's lecture at the Collège de France on neo-liberal governmentality. *Economy and Society*, *30*(2), 190–207. http://dx.doi.org/10.1080/03085140120042271

Lemke, T. (2002). Foucault, governmentality, and critique. *Rethinking Marxism*, *14*(3), 49–64. http://dx.doi.org/10.1080/089356902101242288

Lev, D.S. (1985). Colonial law and the genesis of the Indonesian state. *Indonesia*, *40*(1), 57–74. http://dx.doi.org/10.2307/3350875

Levang, P. (1997). From rags to riches in Sumatra: How peasants shifted from food self-sufficiency to market-oriented tree crops in six years. *Bulletin of Concerned Asian Scholars*, *29*(2), 18–30.

Li, T.M. (2000). Articulating indigenous identity in Indonesia: Resource politics and the tribal slot. *Comparative Studies in Society and History*, *42*(1), 149–79. http://dx.doi.org/10.1017/S0010417500002632

Li, T.M. (2001). Masyarakat Adat, difference, and the limits of recognition in Indonesia's forest zone. *Modern Asian Studies*, *35*(3), 645–76. http://dx.doi.org/10.1017/S0026749X01003067

Li, T.M. (2007a). Practices of assemblage and community forest management. *Economy and Society*, *36*(2), 264–94.

Li, T.M. (2007b). *The will to improve: Governmentality, development, and the practice of politics*. Durham, NC: Duke University Press. http://dx.doi.org/10.1215/9780822389781.

Li, T.M. (2010). Indigeneity, capitalism, and the management of dispossession. *Current Anthropology*, *51*(3), 385–414. http://dx.doi.org/10.1086/651942

Mamdani, M. (1996). *Citizen and subject: Contemporary Africa and the legacy of late colonialism*. Princeton: Princeton University Press.

McCarthy, J. (2000). The changing regime: Forest property and *Reformasi* in Indonesia. *Development and Change*, *31*(1), 91–129. http://dx.doi.org/10.1111/1467-7660.00148

McGregor, A. (2010). Green and REDD? Towards a political ecology of deforestation in Aceh, Indonesia. *Human Geographies, 3*(2), 21–34.

Mehta, U. (1997). Liberal strategies of exclusion. In F. Cooper & L. Stoler (Eds), *Tensions of empire: Colonial cultures in a bourgeois world* (59–86). Berkeley: University of California Press. http://dx.doi.org/10.1525/california/9780520205406.003.0002.

Moore, D. (2005). *Suffering for Territory: Race, Place, and Power in Zimbabwe.* Durham, NC: Duke University Press. http://dx.doi.org/10.1215/9780822387329.

Nally, D. (2008). "That coming storm": The Irish Poor Law, colonial biopolitics, and the Great Famine. *Annals of the Association of American Geographers, 98*(3), 714–41. http://dx.doi.org/10.1080/00045600802118426

Nelson, R.A., Cramb, R.A., Menz, K.M., & Mamicpic, M.A. (1998). Cost-benefit analysis of alternative forms of hedgerow intercropping in the Philippine uplands. *Agroforestry Systems, 39*(3), 241–62. http://dx.doi.org/10.1023/A:1005953032133

Pels, P., & Skalemink, O. (1999). *Colonial subjects: Essays on the practical history of anthropology.* Ann Arbor: University of Michigan Press.

Peluso, N.L. (1992). *Rich forests, poor people: Resource control and resistance in Java.* Berkeley: University of California Press. http://dx.doi.org/10.1525/california/9780520073777.001.0001.

Procacci, G. (1991). Social economy and the government of poverty. In G. Burchell, C. Gordon, & P. Miller (Eds), *The Foucault effect: Studies in governmentality* (151–68). Chicago: University of Chicago Press.

Read, J. (2009). A genealogy of Homo-Economicus: Neoliberalism and the production of subjectivity. *Foucault Studies, 6*(1), 25–36.

Rose, N. (1999). *Powers of freedom: Reframing political thought.* Cambridge: Cambridge University Press. http://dx.doi.org/10.1017/CBO9780511488856.

Sonius, H.W.J. (1981). Introduction. In J.F. Holleman (Ed.), *Van Vollenhoven on Indonesian Adat Law* (29–57). The Hague: Martinus Nijhoff.

Thomas, N. (1994). *Colonialism's Culture: Anthropology, Travel and Government.* Princeton: Princeton University Press.

Valverde, M. (1996). "Despotism" and ethical liberal governance. *Economy and Society, 25*(3), 357–72. http://dx.doi.org/10.1080/03085149600000019

Walker, A. (2004). Seeing farmers for the trees: Community forestry and the arborealisation of agriculture in northern Thailand. *Asia Pacific Viewpoint, 45*(3), 311–24. http://dx.doi.org/10.1111/j.1467-8373.2004.00250.x

Warren, C. (2005). Mapping common futures: Customary communities, NGOs and the state in Indonesia's reform era. *Development and Change, 36*(1), 49–73. http://dx.doi.org/10.1111/j.0012-155X.2005.00402.x

Yap, E. (1998). The environment and local initiatives in southern Negroes. In P. Hirsch & C. Warren (Eds), *The politics of environment in southeast Asia: Resources and resistance* (281–302). New York: Routledge.

PART 2

Neoliberal Technologies and Politics

5 Governing Emergent Technologies: Nanopower and Nanopolitics. An Ethnographic Approach

ROB SHIELDS

Nanotechnology has been heavily hyped through the turn of the millennium. The term refers to the technologies necessary to visualize, design, and build at nanoscales (10^{-9} or a billionth of a metre)[1] and to scientific pursuits that examine quantum behaviours of particles and materials on this scale. Like "macro" or "micro," "nano" is a scale of becoming with its own properties. It is qualitatively different, with a complex relationship to other scales of any given physical element; it is not just quantitatively smaller. Nanoscaled entities are not visible but are sensed haptically, through tactile technologies such as scanning-tunnelling electron microscopy. They are not tangible but exist on the atomic scale as part mass, part energy and thus are virtualities to be objectified through lab technologies and practices.[2] Nanotechnology promises to yield massive economic benefits as well as to serve as a pervasive, powerful platform for reconceiving the creation of products and solutions, both large and small. The latter include diagnostic products (from genetics and "lab on a chip" devices to simple colour-change indicators of health status – e.g., pregnancy and tuberculosis) and solutions for climate change and the bioremediation of wastes. Both a child of and a vehicle for neoliberal forms of research organization, nanotechnology itself is a laboratory for reorganizing the production of knowledge.

This chapter investigates the relationship between innovation in emerging technologies and neoliberalism as it is implemented in the case of the National Institute for Nanotechnology (NINT) on the campus of the University of Alberta in Edmonton, Canada.[3] Neoliberalism in this context is defined as a series of critiques of governments' abilities to understand and regulate the economy and a valorization of the ability of communities to self-govern (Hayek, 1945). I reflect on the case of NINT using ethnographic and geographical approaches, using narrative to capture the complexity of a local place. This goes beyond local socio-economic conditions in order to map the "microphysics" of co-evolving cultural repertoires and community institutions within which neoliberal science policy and programs are deployed. This chapter draws mainly on Hayek's presentation of neoliberalism as well as the understandings presented by a diverse group of participants in a participatory study. Drawing on and critiquing Mitchell Dean's four axes of governance, governmentality is given a fifth affective dimension under which to examine the interaction between neoliberal forms of capitalism and research in emergent sciences, as recounted in the narratives of participants in our ethnographic study.[4] There were a total of 104 participants, who represented the local public, experts, and various policy and industry actors (see below). Finally, I hypothesize that nanotechnology may initiate a shift from a biopolitical to "nanopolitical" governmentality that requires a fundamental recasting of what we understand as the governmental rationalities of neoliberalism, its subjects, and its objects.

Nanotechnology emerged from a disparate set of micro-electronics, atomic and quantum applications as a new, synthetic field based on access to the new, shared facilities, labs, and tools such as scanning-tunnelling electron microscopes. Unlike the Cold War stress on engineering and the physical sciences, nanotechnology included biosciences in the US National Nanotechnology Strategy (NNI), a competitive response to global leads by Japan and Germany in synthesizing carbon nanotubes (see Crawford, 1991; JTEC, 1994).

Nanotech has been also seen as echoing neoliberal notions of autonomy and self-organization, "In much the same way that [Hayek's] neoliberal theory regards markets as self-regulating mechanisms that instantiate a higher degree of rationality than any individual would be capable of" (Gelfert, 2012: 160). It links the market as a spontaneous order with an ideal of autonomous voluntary communities (Hayek, 1976: 151). "In the discourse on nanotechnology, 'self-organization' (with or without self-replication) functions as the analogue to 'spontaneous order'" (Gelfert, 2012: 160–1) as nanobots evolve adaptively and autonomously, supporting small, "self-sufficient communities" "without bureaucracies or large factories" (Drexler, 1990: 235). The association of nanotechnology and neoliberalism is a coincidence, but their parallels can be more critically examined in the context of changing institutions and nanoscience projects that search for commercializable knowledge.

Nanotech in a Northern Climate

Edmonton and the Alberta Capital Region (population about 1.3 million) is located in the Prairies at the centre of the province of Alberta, making it the most northerly large Canadian city. As such, it is currently a hub for the northwest with an oil and gas services economy. However, Edmonton was an Aboriginal meeting place at a river crossing and a Hudson's Bay Company trading outpost from the late 1700s. Furs were Edmonton's earliest commodity. At first they were shipped down the river that winds its way towards Hudson Bay; later, Edmonton was the northern terminus of late nineteenth-century railroads and the capital of a new province from 1905. The place has long been seized by patriarchal, colonial, and global empires and participates globally in contemporary military adventures. As such it is one of those bellwethers of economic change at both the local and the global levels.

It has remained unexpectedly prosperous through recent economic and oil-price downturns, having a less well-known university-based research environment, military base, and a history of starting up mass-market retail chains and early role-playing videogames. Edmonton is known for hockey teams, inventing the mega-mall in the early 1980s (Hopkins, 1991), and pioneering municipal recycling and waste diversion. Carol Greenhouse argues that neoliberalism encourages these mass forms of mechanical solidarity over and against the interdependent organic solidarity of more specialized twentieth-century metropolitan economies, changing relationships and raising ethnographic questions concerning communities (2011: 3; see also Comaroff and Comaroff, 2001).

In 2002 the provincial and federal governments funded the creation of a US$200 million National Institute for Nanotechnology (NINT) with hopes that it would foster a more diversified economy in Alberta (AITF, 2007). The residents of Edmonton (from 2004 to 2014 one of the top four local economies in North America, by regional GDP) found themselves a would-be engine of a new source of wealth in a globalized economy, perhaps "branding" the city region as the seat of nanotech products "when the oil runs out." However, more familiar with mass-market sports, consumption, and production forms, our respondents doubted that NINT's elite efforts were synchronized with local needs (see quotations below).

NINT has proven successful in the academic world, but the other "spin-off" benefits such as jobs and urban renewal have been slower to emerge and are less easy to assess. How can the dream of nanotechnology innovation be translated into the reality of enhanced regional development and direct benefits for the local community? How does one make innovation driven by economic goals locally relevant to the needs and desires of citizens? Research on commercialization (Beaudry & Schiffauerova, 2011) and academic publications (Hu, Carly, & Li, 2012) identify Edmonton as having a medium-sized concentration of nanotechnology companies within Canada, albeit less integrated into the global market compared with Ottawa, Toronto, or Montreal.

The establishment of NINT has been part of an ambition to create a regional industrial concentration of nanotech firms. These "industrial clusters" are a spatial construct that approach a reflexive sense of "economic place" in that an agglomeration of activity in or supporting a given sector is understood to exist (Porter, 2000; Wolfe & Gertler, 2004) even though no clear or stable boundary can be drawn. Nor do clusters have a minimum density or size. These are spatial expressions of concentrations of capital and talent sufficient to determine the direction of global markets and to become centres of decision making as well as production and thus sites of economic power. They refer to how economic activity is cast in a spatial form or to "social spatialization."

The space or site in which science occurs matters. The Edmonton city region is seen by many as lacking charm: an "ugly" place planned around the car, despite the initial quality of its Prairie environment. It is a region where high-tech industry doesn't quite belong (Gow & Sandy, 2007). It is important not to underestimate the impact on both residents and investors of media representations of place and of rising or falling global economics and military coalitions. These discourses add to, or negate, a sense of sustainability into the future, especially given ambivalent place images (Shields, 1991) and Edmonton's negative place myth of a sprawling, blue-collar, winter city. Despite its distance from other centres, both global oil prices and global politics, including the invasions of Afghanistan and Iraq/Syria, impacted the city region viscerally during the five years of our investigations.

Such spatializations cast these places as summits within a wider topology of "places-for-this and places-for-that." Ferguson & Gupta (2002) argue that neoliberalism entails further changes in the spatializations, tying state authority to the territorial scale of the nation-state. Neoliberal transnational ties change the "vertical" understanding of how localities are "encompassed" by and exist "within" nation-states amid cross-cutting transnational, national, and regional scales of organization, governance, and authority. Analysis of places and clusters is often further complicated by the place becoming a metaphor for the activities it hosts (Shields, 1991). The rhetoric of clusters thus tends to attract and concentrate capital and talent as a type of self-fulfilling prophecy. However, one of the chicken-and-egg problems of the literature on clusters is a lack of critique of the extent to which it has relied on an unexamined spatialization that overestimates the power of the abstractions and theoretical representations of places produced by urban and regional planning and economic management professionals to frame our imagination of what places can be (i.e., as places or "spaces of representation") against actual, everyday practices (ibid.).

Ethnographic Approach to Governmentality

Governmentality is the set of techniques and tools by which activities are governed, where conduct is shaped by a cultural system of values and knowledge (see Pyykkönen, 2007). It is expressed in social spatializations that serve as proxies for the governance of affect (Davidson, Park, & Shields, 2011: 113), catering to and legitimating seriousness in centres of production and decision making, and levity and play in liminal zones and ludic margins (Shields, 1991). As a simplified example, consider that Edmonton, as a site of production and long, double-shift work hours, is connected by many direct flights to its spatial alter-egos, Las Vegas and Palm Springs: sites of relaxation, play, consumption, sun, and warmth (Davidson et al., 2011: 113). Governance also changes in time insofar as it is structured as a historical configuration of knowledges of government, governing technologies, and conceptions of the subject (Foucault, 2008). At any given time, each of these modes of governmentality seems completely commonsensical when taken on its own terms. Neoliberal political rationalities reconfigure the relationship between the state and citizens to govern individuals indirectly as customers who are responsible for their own choices and well-being, rather than having an expectation of guarantees of their quality of life underwritten by the state as under the liberal welfare state of the twentieth century (ibid.; Joseph, 2012; Rose & Miller, 1992).

The discussion of governmentality will be organized using Dean's (2009: 33ff) description. Dean argues that governmentalities are actualized on four axes that operate in analyses of discourse and practices:

- Its objects of governance, both actual and possible, such as risk: What do we seek to act upon?
- Technologies, means and systems of governing including rationalizations, and informational technologies which reveal or make knowable processes, such as the statistical surveying of populations: How do we govern this substance (cf. Dean 1996)?
- The roles, responsibilities, rights, and duties within institutions that are internalized by actors: Who are we when we are governed in such a manner?
- The ideals or principles which serve as ethical objectives and social goals: Why do we govern or why are we governed?

To these axes we will see that a fifth, *affective* dimension must be added to the analysis of governmentality. Lacking it, such models do not capture sentiment, the swings of individual outlook and collective mood, which are important motivational aspects of everyday life that are poorly institutionalized and often figure in social analyses only as suppressed elements of individual psychology, such as models of professionalism at work and civility in public life.

It is also important to note that the discourses that are often the focus of studies of governmentality derive their significance from being mutually correlated rather than through absolute denotation. Within these studies the self-evident character of rationalities of governing has to be subjected to an immanent critique to show the artificial quality and conditions of possibility of any mode of governmentality. Our research addresses criticisms that the link between the structuralist analysis that underpins studies of governmentality and empirical cases is weak (Marrtila, 2013). How does governmentality, as a sustained configuration of heterogeneous but consistently effective statements, practices, and institutional norms, relate to the selection of cases and their analyses? The dynamics of nanotech innovation in the Edmonton region were examined using a qualitative, case study approach (Mason, 2002) and grounded in a "public research" model (see Patchett & Shields, 2012) that drew its research questions from community workshops that mobilized public curiosity.

Are studies of "governmentality" captive to second-order observation issues (Fuchs, 2001) or to a post hoc ergo propter hoc bias (by which the analysis reveals just what theory highlights by fitting empirical data into categories established in advance)? To respond to such critiques of bias, Bourdieu &Wacquant (1992) propose a self-reflexive approach to social research generally. This methodology recognizes that social reality is mediated by the discourses of respondents and other sources. In our research, we are therefore cognizant that we are dealing with respondents' stories about the world, not the world per se, and divergences, omissions, what is stressed, and what is not talked about are of interest. From 2010 to 2015 in partnership with Kevin Jones and Nils Petersen, we conducted focus groups followed by a weekend "Citizen Summit" (Irwin, 1995; Shields, Jones, & Petersen, 2013) on nanotechnology, participatory presentations to the public, and tours of key local innovation sites (Davies, Selin, Gano, & Pereira, 2013; Jones, Shields, & Lord, 2015), from which we gathered feedback in the form of audio transcripts, participatory photography, and reflections during events hosted by Edmonton's Telus World of Science museum.

All of these complementary forms and methods extend the opportunities for capturing different perspectives and meaning-making about the complex of policies, opportunities, and local needs by those directly involved over time. In addition, we observed participants responding to social science expertise and theory – not in an anecdotal tone but with a sense of evaluation of local-global conditions. Global neoliberal discourses appear as only one element to be weighed.

The problem of communication has become a major day-to-day limit on collaboration and coordination:

> But having been [at NINT] for a while, talking to people, one of the big issues that not a lot gets out I think is based on intellectual property issues. The government's goal, or as far as I can tell, NRC's goal, is that tangible products come out of the research at NINT. They want industry to invest, and so I was involved last week in working on a project that was going to look at knowledge mobilization from one of the research groups at NINT. And he's partnered with industry and the CEO of that company which is privately held. He was so worried about – he didn't actually want us to have access to any of the information because he said "people can infer, you may not even be intending, you can talk about general stuff but if you're in the industry then you can infer x, y, and z. And the next thing I know my intellectual property is gone." So we can't really engage in this project because your tangible product might be compromised in some way. (Focus group participant, 2011)

Over and above Dean's objects, roles, and ideals of governance (above), the *affect* of anxiety and fear in the above quotation from a

respondent is an essential but overlooked aspect of neoliberal marketization. A rollercoaster of emotions that cycle between heights of
empowerment and ambition followed by melancholic dives into dis-
empowerment and frustration accompanies the rational acquisition
and processing of information. This is true even in rational technical
research and cautious investment practice. Perhaps the most important
contrast between what our respondents report and theories of neoliberal governmentality is this role of affect and other intangible virtualities, including trust and identity (see quotations below; Shields,
2003). Lived neoliberalisms are heady ecologies of affect, rashness, and
confusion.

Neoliberalism, Knowledge, and Research

Lack of understanding of action, sentiment, and choices is at the heart
of the neoliberal critique of welfare economies. As neoliberal rationalities have become dominant, state research and development, a legacy of the military investment in science and technology during the
World War II and the Cold War years, has been selectively privatized
and de-funded. On the one hand this process has been understood as a
reduction of state technological capability in support of regulation and
forensic testing of products and processes. On the other hand, it has
constituted a withdrawal of the public from most exploratory science
or technological projects. With Hayek as the best known historical proponent from the 1940s, neoliberal rationalities have advocated the creation of markets in all systems, even the environment and climate (Oels,
2005). Although the parameters of these rationalities may vary, they
tend to lead to the same outcome: the reduction or elimination of funding and the closure of agencies or the restriction of their activities in
favour of market-based responses to opportunity and challenge. Examples include both funding reductions for academic research and the
reduction of national research agencies such as NASA or the National
Research Council of Canada.

What remains as a focus of public funding is the formation of scientific expertise in support of the high-tech sector and, increasingly in
the 2000s, in sectors that support military and security technologies. In
turn, these sectors draw on frontiers of science that are not commercial
endeavours but accrue prestige for the state and underpin participation in international collaborations that could be understood as a form
of scientific diplomacy. This approach has favoured nanotechnology

and other highbrow sciences at the frontiers of technical capacity and theoretical insight (e.g., particle physics and theoretical cosmology) or scientific missions that legitimize and actualize contested claims of sovereignty such as in the Arctic (e.g., CHARS [Canadian High Arctic Research Station] in Cambridge Bay, Nunavut).

While the variations of the theory are similar insofar as they are distinct from a simple laissez-faire market capitalism (Foucault, 2008). Despite critiques in academia of business influence, values, and management techniques since the time of Thorsten Veblen (the important early modern social analyst of the governance of education and training [1918]), Canadian federal policies have been used to encourage private investment in science and partnerships between universities and industry, in part by eliminating government regulatory labs and decreasing public funding for academic research (see Brownlee, 2014; Buchbinder & Newson,1990; Lave, Mirowski, & Randalls, 2010: 661–2). The explicit goal of Canadian science and technology policy is to increase commercialization and to direct research towards international collaborations in areas of strategic importance to the Canadian economy (Industry Canada, 2014). However, an important brake on shifting to the stereotype of the neoliberal, business-minded, and corporately tied university in Canada is that educational institutions are not vested nationally but regionally within the jurisdiction of each provincial government (Schuetze & Bruneau, 2004).

As neoliberal rationalities have taken hold, information and knowledge have been conflated and treated as commercial goods (Drake, 2011). This shift links academics and universities as knowledge producers and knowledge institutions to commercial interests (Brown, 2000) with intellectual property (IP) as the central "object" produced and monetized in the form of patents and royalties managed via Material Transfer Agreements (MTA). Patenting has in turn become as important to science careers and evaluation as publications and citation rates, leading to a rise in "vanity patents" and a focus on searching for patentable knowledge, rather than, for example, laws of nature (Lave et al., 2010).

However, this is hardly a one-way street. These changes have been implemented in the context of inherited political and economic rationalities and configurations and local socio-economic conditions (see Dorow, Howard, Li, and Mitchell and Lizotte in this volume). For example, struggles between regional elites or by neoliberal political ideologists against the fiscal autonomy of scientific and medical elites that depend on state financing (in the case of all nationally funded medical systems) inflect implementation. Struggles to brand policies with the identities of the political party in power or to favour particular economic lobbies or industries transform neoliberalism into a kaleidoscope of actual policies and political rationalities.

Nanotechnology, Neoliberalism, and Place

A significant problem that has emerged from the federal government's approach to nanotechnology and other emergent sciences is that top-down models of innovation investments fail to recognize that development is difficult to direct (Bair, 2008; Lorentzen, 2008; Varlander, 2007).[5] They ignore what Elias referred to as the "figuration" in place with its local and institutionalized divisions between cliques, insiders, and outsiders (Elias & Scotson, 1994). Approaches such as that of Elias and Scotson can further extend governmentality from general theory to the specific affects, embodiments, routines, and "figurations" of neoliberalism. It is difficult even for insiders to cognitively map the local network, which also has important "players" who may be international partners elsewhere, or coalitions of clusters that constitute a transnational scale of organization folded into the local insider enclave of the would-be nanotech "cluster." Yet the local site is significant: in the case of nanotechnology, exceptional facilities with "clean rooms," and even a silent seismic and radio-frequency environment, mean that facilities for scanning-tunnelling microscopes cannot be situated just anywhere and require extensive technical staff to maintain this exacting environment. Lacking a broader view, respondents often focus on the objects of governance: "I've worked with a consultant who previously worked for [company] ... And he honestly didn't know about all of the facilities, nanotechnology facilities and supporting infrastructure. So I was shocked because he didn't know, he was shocked because he didn't know, and he's in the industry" (Focus group participant, 2011).[6]

Until almost a decade after NINT opened, there was little investment in policy infrastructure to promote commercialization beyond NINT, leaving such endeavours to private investors and a university-municipality partnership in a commercialization incubator, TEC Edmonton. This problem of the connecting tissue from innovation to products globally in use was often raised in focus groups, tours of innovation sites, and at the Citizen Summit (Shields et al., 2013):

Okay. So that piece is – there's a lot of nanomedical companies and they need to get in, and we can't even sell it here. They have to go always somewhere else. So that's what you can help with. (Focus group participant, 2013)

Ninety per cent of our products that we make are sent outside of Canada, for someone else. We have a company that's asking us to make this sensor to find oil underground. It's great but it goes to Saudi Arabia; it goes to Mexico. We have the largest oil deposit in the world a couple of hours north. We do nothing in that space and we don't have customers that want us to do anything in that space just yet. That's part of it, is that they don't even know that we exist. (Focus group participant, 2011)

The spatial complexity is significant. The discussions around nanotechnology and the exploration of the weave of companies, training programs, incubators, NINT, governments, and global corporate interests shifted to a focus on innovation and the institutional isolation of some of the local key players (Sluggett, Shields, & Jones, 2015). Elsewhere we argue that research fora such as the Citizen Summit allowed "difficult" discussions on local futures to take place (Shields et al., 2013):

P1: Are there tensions then between not just the science, but between the university and the overall city, in terms of where [nanotechnology] fits and what that relationship is?
P2: I think there is just no relationship.
P3: There is no relationship.
P2: There's not tension, it's just non-existent. (Focus group participants, 2011)

There has never been, for example, a strong identification between nanotechnology and the city region: no "nano-Edmonton," or "nano-north," especially in comparison with the much-hyped image of the hydrocarbon resource economy which experienced both boom (2004–8) and bust (2008–11), partial recovery (2011–14), and another slump (2014–15) in oil prices driven not by economic supply but by international economic warfare and competition as US shale-oil flooded the market, followed by Saudi exports.

In group discussions participants struggled with the location and spatialization of Edmonton and the power of stereotypes to undermine innovation. They raised, again and again, questions of community, civic and place identity, and how they relate to each other in a complex trialectic:

P1: Thinking about our unique context of how far north we are and how popular we are compared to other towns or cities this far north and like that's almost like a human-centred design approach for innovation is like thinking about your context and then going forward with new ideas based on that too.
F: So our northness is an advantage in innovating in some things, is that right?
P1: Yeah, or like creating solutions or ideas that are really based around that human context too for our city.
P2: It sort of forces us to adopt that mind frame of innovation because we don't have a choice.
F: That's interesting because in our earlier research a lot of people said we can't really be an innovative city because we're far too much of a winter city, we can't attract people, people don't want to live here, too far away from markets.
P3: And I think those are – we use those challenges – it's about if that's the local context you need to design or innovate or like plan to that context and I think that's what can be overcome. You can overcome it and plan through it and recognize it as opposed to adopting what someone would do in say, Florida, because clearly we're not Florida. (Focus group participants, 2013)

Place therefore blurs with and challenges us to rethink Dean's second axis of "means" as a suite of capabilities and affordances that contribute to an ecology that is both concrete (objects) and virtual (a set of elements including place myths and identities, an atmosphere or milieu). They also allude to challenges or "tests" that validate these capabilities (cf. Boltanski & Thévenot, 2006). It is important to note the way that ambition as an affect is tied to place through boosterism (see next quote). As an example of the neoliberal governance of affect through spatialization, citations of place accompany calls for the investment of affect in a collective project.

The question of scale is central not only to nanotechnology but to neoliberalism because of its incubation and promulgation in international institutions such as the International Monetary Fund, and the explicit pairing of global markets with local productivity, identifying national economic and import policies in between as the barrier to harnessing the two extremes together (Ferguson & Gupta, 2002; Greenhouse, 2011; Larner, 2001).

However, the shift to discuss innovation was also a moment in which the nanoscale and the challenges of new materials and objects were reinserted into a comfortable dominant discourse of private entrepreneurial innovation. At the level of detail that was being discussed a different participatory research process, such as responding to different future scenarios, would have been necessary to imagine and explore alternative figurations and spatializations of innovation. Such alternatives are nonetheless clearly evident in the implications of nanotechnology, which include not only neoliberal narratives of knowledge, innovation, and market success but also disruptive innovation that completely recasts these linkages.

Essential to neoliberal rationalities is the biopolitical production of "responsibilized" forms of individuation and subjectivity (Collier, 2009: 81; Foucault, 2007, 2008; Rose, 1999) that direct attention to individual reflexivity, choice, and risk taking (Jones & Irwin, 2012). Respondents often repeated the formula of individual responsibility and industry. This is not merely a matter of internalizing responsibilization but integrating it with the collective interactions, routines, and projects of the city region as both an aspiration and present context (see also Degen, 2003). Nanotechnology spills over into neighbourhood revitalization as part of an enabling rhetoric of the affective "grandeur" of place and community:

> I see Edmonton in a new light; my perspective of the city has shifted from grim unease to that of cautious optimism for our future –provided we can work together to foster the kind of environment that promotes innovation and the creativity necessary for large-scale change, social evolution and advancement to occur. Participating in these kinds of activities is what will drive our collective goals to fruition, and, after experiencing this phenomenally eye-opening tour, I would encourage Edmontonians to engage in constructive, forward-thinking conversation with one another about the nature of our neighbourhoods, the progression of our local projects and developments, where our decisions today will lead us in the long term, and whether that vision of the future promotes us as an adaptable and sustainably focused hub of innovation. (Focus group participant, 2013)

Analysts have argued that the entrepreneur as a figure embodies these changes and appears frequently in popular and media discourses as a sort of hero figure or ideal (see Miller & Rose, 1990) – an idea that appears in the popularity of the epithet "social entrepreneur" and the notion of the "academic entrepreneur" and is reflected in the changing bases of evaluation of success in academia. This concept was also reflected in our participants' comments, which identify the ideal entrepreneur as alienated from institutions such as incubators and commercialization programs (a situation that is perhaps changing over time with the revision of this ideal notion of the rugged individualistic entrepreneur):

> P1: So I think that part of this collaboration is just to get the companies together. I mean it's great that they're bumping up in these facilities, but many companies I know won't go near those kind of facilities, they won't go near TEC Edmonton.
>
> P2: Absolutely, yeah.
>
> P1: They think very differently, and it has to be entrepreneur-to-entrepreneur because that's the only kind of people they trust. And I'm sorry, I know you guys are – I'm saying it to your face – but they will only trust entrepreneurs who are like them. And it's almost like they're part of a knighthood [laughs] and you have to be one, you can't fake it, and that's sort of how they think. So we need more opportunities for collaboration, you're absolutely right, but part of who they are is it has to be entrepreneur-to-entrepreneur. (Focus group participants, 2013)

These comments – with trust at their centre – suggest that the institutional form of nanoscience and technology expresses a convergence not only around shared tools but also of shared projects and ideals that link, for example, computer visualization (of nanoscale phenomena) to physics, and applied chemistry and materials science (such as new lightweight structures and materials) to medicine and bioengineering to create profitable products such as prostheses and medical devices and coalitions of entrepreneurial subjects. The institutional isolation noted earlier is evoked as a need for "more opportunities for collaboration" but with an important caveat of who is admitted. An affective belonging, not just place based, propinquitous community is required, challenging the neoliberal citations of place. What is significant is this struggle over the power of the local as binding people and public and private actors around projects such as a nanocluster.

Even if supported by medical insurance or state healthcare, these products are intended for global markets that have been opened up under neoliberal trade agreements. There is no simple hegemony of private interests or of global corporations, but both the purpose and the bindings of scientific research activities have changed in response to demands such as those of the Canadian Business Council on National Issues and the Corporate Higher Education Forum (BCNI, 1993), giving rise to debate and contestation around identities, powers, and agendas to include training and private funding in research. Science is produced for particular clients and market interests, while impacts and issues born by others may be left as undisclosed or unresearched "white areas" on the "map" of science.

As more technologies implicate bodies, cultural identities, and the environment and circulate around the world under neoliberal trade arrangements, science has become implicated in a wide array of social movements of the right and the left, from large professionalized national networks to small under-resourced community groups (Moore, Kleinman, Hess, & Frickel, 2011: 528). This association has both benefited and harmed those who deal with the products and wastes of nanotechnologies (Pellow, 2007), thereby altering the relationship of science to trust and authority and has brought others into the governance of science and technology, a process referred to as "epistemic modernization" (Moore, 2008; Moore et al., 2011). Corresponding to the global standardization of markets, the locus and arenas of governance have moved to the international scale and into the hands of technical experts. Rather than being values based, regulation has become a norm, which has limited the scope for government regulations (e.g., The International Standards Organization and other coordinating bodies; see also Della Porta & Tarrow, 2005; Gibbons, Limoges, Nowotny, et al., 1994).

Nanotechnology and Neoliberal Research

In Von Humboldt's organizing vision of the modern multi-subject university divided between the arts and sciences, research and teaching went hand-in-hand with a pedagogy for creating critical citizens. In the establishment of NINT, advanced research was removed from the "learning spaces" of the university to a secured state institution that barred the idly curious and admitted only select members of research teams. This separation had a negative impact on the institute's local image, but reflects the priority given to commercializable research for global markets over teaching in the neoliberal university (Lambert, Parker, & Neary, 2007; Larner & Le Heron, 2002): "There is an intimate

connection between the neoliberal recasting of the market as an information processor, and the growth of the conviction that knowledge
should be commodified. This connection seems all the stronger when
one considers that, as several recent studies have pointed out, for the
vast majority of universities patenting has been a losing financial proposition" (Lave et al., 2010: 666).

The withdrawal of public funding and limits on the ability of universities in many jurisdictions to charge the actual cost of tuition create a
tension between the need to meet the market imperative of increasing
commercial sponsorship while allowing research to remain value free
and part of an open public sphere of science. The result is the proliferation of contradictions. What is accepted as good research becomes
contested and adjudicated not only through peer review and other traditional mechanisms of science but by markets and quantifiable practical implementation: "The character of science is changing as privatized
management shifts the sources and quantities of funding, organization
of research and teaching, and the intellectual and commercial status of
knowledge claims" (ibid.: 669).

To this changing institutional context, nanotechnology research adds
a further twist. It encourages the convergence of disciplines around
shared tools and facilities that would otherwise be out of reach of most
universities and commercial establishments. Given this expense, nano-
characterization and research labs have been justified as *workshops*
producing not only discoveries but prototypes and new hybrid instru-
ments and tools for the benefit of multiple disciplines. It is not unusual
to find an osteopath co-funding and sharing a lab with a civil engineer
interested in the properties of wood because both use the same scan-
ning equipment.

Nanopower and Nanopolitics

The objective of Foucault's "biopower" was to show how the biological
details of humans became the object of political strategy, that is, a form
of governmentality. Similarly, I could suggest *nanopower*, in which the
quantum possibilities of all existence become a mode of governmentality of all matter, things, and subjects: *nanopolitics*. This becomes both an
ideal and a practice that reconfigures the arrangement of disciplines,
institutions, industrial sectors, and Enlightenment divisions such as
those that exist between the body and its environment, between subject and object, between static solid and dynamic fluids, and between

matter and energy. Although this analogy may seem far-fetched, nanoscale processes and objects already are present in everyday life (e.g., CD and DVD recording processes and air fresheners, respectively) and have realigned disciplines and university-industry relationships, as argued above.

Drawing on our participatory research, respondents – mostly highly educated – identified these shifts without reference to governmentality, neoliberal economics, or ideologies. However, familiar tropes, such as globalization, risk, entrepreneurialism, and new forms of property, often appeared. Participants were ambivalent about familiar neoliberal objects of concern: the role of the state, the autonomy of science and academic institutions from the local economy, and incomplete transitions to the knowledge economy.

How does the local figuration (Elias & Scotson, 1994) around nanotechnology coincide or depart from the axes of governmentality suggested above? All four of Dean's axes shift. In particular, the objects of governance require a more nuanced ontology beyond the division of "actual and possible" (an unexamined material-ideal dualism at the core of most theories of governmentality to date). Nanopower offers an atomic view of the world from processes below the threshold of everyday life that unify biological and physical, bodies and objects, and static things and quantum processes. It is the ultimate "naturalization" of being and the order of the world while at the same time including probabilistic and virtual objects as well as material and abstract entities. The required technologies of governance subsume biotechnologies. The qualities of nano-objects depart from those of the same element at the scale of everyday objects, meaning that its properties can be manipulated (e.g., by changing the energy level of ions, colours change, thus allowing coloured indicators to be used in technologies such as a home pregnancy test). Because of the huge surface area of such fine versus gross everyday particles, they are exceptionally soluble and reactive at room temperature, transforming metallurgy and synthetic operations. At the same time, new forms of risk and opportunity arise, such as unanticipated applications and interactions and challenges around containment and contamination.

A significant shift in the understanding of "manufacture" and "creation" accompanies nanotechnology. Rather than "grinding down," nanotechnology "builds up" from elementary particles and energy states. A simple example is found in the shift from photocopy toner created by grinding carbon, to synthesized, multi-coloured toner with consistent, much finer particles, which has allowed personal laser printers to achieve high resolutions. A similar idea is found in the conception of 3D printing. We have not explored how nanotechnologies intervene in and alienate organisms' practices of reproduction, radically deconstruct sexualities, or how they support a heretofore unimaginable recasting of gender identities and forms of human and other species' embodiment.

Nanotechnology eludes contemporary political forms that have evolved to represent and govern tangible entities. Only when anthropomorphized as "nanorobots" – in other words, as new actors or personae in Dean's third axis – did our respondents interact with the risk and wider social implications of nanotechnology. This result is reflected in a relative lack of affective engagement that we discuss elsewhere (Ghimn & Shields, 2015): while respondents quickly supported the boosterism around NINT and place, they seemed neither articulate nor sure how they *felt* about nano-objects and nanotechnology. A general silence of the humanities on the ideals and ethical objectives of nanotechnology, a lack of counter-movements, and a failure of critical thinkers ensure the development and success of nanopower, but this quietude has slowed the understanding and adoption of nanotechnology because of lack of awareness, perhaps merely postponing other difficult conversations.

Nanotechnology is presented as pervasive, persistent, and powerful. It permeates and displaces more complex totalities such as mind, environment, and climate, proffering "can-do" technical solutions to mega-problems such as depression and anxiety, pollution and climate change. Nanotechnologies require a more nuanced understanding of the real as not only actual (concrete, present) but virtual (ideal-real), where "things" have multiple states (or energy levels) and are co-presences of potentialities and latent states. Nano-objects are controlled neither directly as actual things and actions, nor abstractly through representations, but instead they are managed probabilistically, as actual statistical possibilities (see Shields, 2003). Our respondents reported their development of standards and ethical practices as a bottom-up development of laboratory techniques, not as an ethical project.

Our ethnographic study shows that neoliberalism is being implemented in the geographic context of Edmonton, Alberta, and Canada as a place governed according to overlapping or nested jurisdictions: federal science and technology policy, joint federal and provincial funding of NINT, and the city in partnership with the academic institution to promote commercialization in the form of technology start-ups. These administrations historically have developed distinct governmentalities with their own social and cultural strengths, capacities, and configurations. Our respondents agonized over the gaps in policy and outcomes that resulted from lack of coordination or from conflicts between institutions at these different scales:

> I really wasn't tuned into nanotechnology until a couple of years ago. It was something that was germane to electronics, but it wasn't something that was all that germane to the forest sector. So I don't remember all of the hype and the expectations that were being touted. I would suggest to you that there probably isn't enough said about NINT, the university, and nanotechnology and business spin-offs. I don't think the community is all that aware that it even exists. There's probably more awareness amongst some of the international community than right here in Edmonton ...The average citizen of Edmonton probably doesn't realize that NINT exists at the university. (Focus group participant, 2011)

Our participants were candid about the struggles between academics, the organization of universities around Fordist mass-education of the welfare state, and neoliberal demands for elite socialization of leaders, plus technical vocational training. These difficulties also led to conflicting roles and a tension between the global corporate orientation of NINT and the community's demands for transparency and orientation to local benefits (although they stopped short of demanding a role in governance). They were also willing to reconfigure their city to enhance corporate collaboration at other levels and to erode the separation that marks the university, with its campus and NINT on one side of the river, from downtown, on the other side. This remains an evolving situation, only a decade into an important socio-economic experiment. Our research turned to affect and place as missing registers in the analysis of governmentality, notably the spatialization of place as a mode of governing affect – in Edmonton's case the project to create a nanocluster. In relation to Dean's four axes of governance, our respondents can be read to suggest:

- the significance of a new scale and range of material and virtual objects of governance we act upon;
- new means, technologies, and systematic approaches in knowledge and economic activity in which the haptic and other sensing technologies are linked to the visible through computer-aided

visualization; and the prominence of affects such as trust, a sense of community, and place as logics of social assembly and mediators of interaction;

- neoliberal roles include the entrepreneur, the researcher, the nanoro-bot, and the citizen, but there is a less explicit understanding of the deconstruction of bodies into basic biochemical processes by nanotechnology;
- challenges lie in ethical judgment and an absence of debate: risk is to be managed technically but the goals and ideal ends of nanotech-nology are not considered.

Nanopolitics takes seriously the claims that a pervasive, persistent, and powerful new basis for technology and industry has been intro-duced through visualizing, controlling, and designing at the nanoscale that includes both the organic and the inorganic. This process intro-duces new objects of governmentality while synchronizing with and deepening neoliberal and biopolitical modes. It is produced within a neoliberal logic that applies science as innovation, while creating new entities and actualizing new capabilities in the ordinary physical ele-ments. In so doing, it changes how we understand the real and the possible, fundamentally altering the biopolitical and governmental rationalities of neoliberalism.

ACKNOWLEDGMENTS

This chapter has benefited from the very helpful comments of the editors and Stephanie Bailey as well as the input of participants in research that was part of a Social Sciences and Humanities Research Council of Canada project with Co-Principal Investigator Dr Kevin Jones and Professor Nils Petersen, "Nano-technology and the Alberta Capital Region – A Case Study in Integrating Com-munities, Innovation and Development" in and around Edmonton, Canada, 2010–14 (see Corner & Pidgeon 2012; Joly & Kaufmann 2008). Our team also included Dr Yun-Csang Ghimn and collaborators at other institutions: Profes-sors Jack Stilgoe, Alan Irwin, Cynthia Selin, David Guston, and others. Michael Granzow, Katie Herzog, Bryan Sluggett, and others examined specific aspects of the dynamics of innovation in the region. We greatly benefited from the as-sistance and funding of Faculty of Extension City Region Studies Centre and partners, including AITF (Alberta Innovates Technology Futures); the City of Edmonton Sustainable Development; TEC Edmonton; National Institute of

Nanotechnology; Northern Alberta Institute of Technology; Telus World of Science Museum, Edmonton; and the Henry Marshall Tory Chair Fund.

NOTES

1 One way to visualize 10^{-9} is to imagine that the ratios of scale between a gold-nanoparticle and an everyday grapefruit is the same as between the grapefruit and the planet Earth. Further materials to understand the nanoscale and nanotechnologies are available at http://www.crsc.ualberta.ca.

2 In this chapter, terms such as real, concrete, or the "actually real"; the possible abstraction or the "ideally possible"; virtual, potential, or the "ideally real" refer to the four-part ontology first presented in Shields (2003) and the play between these categories. Proust referred to virtualities such as memory as "real but not actual, ideal but not abstract," in effect setting up a table of the ontological categories: virtual, concrete, probable, and abstract. Nano-objects are not tangible, concrete entities but atomic energy states that blur the conventions of materialism.

3 We conducted focus groups followed by a weekend "Citizen Summit" (Shields, Jones, & Petersen, 2013) on nanotechnology, participatory presentations to the public, and tours of key innovation sites from which we gathered feedback in the form of audio transcripts, participatory photography, and reflection in public events hosted by Edmonton's Telus World of Science museum (Jones, Petersen, & Ghimn, 2013). Participants in our study came from a wide range of backgrounds, including the university, commercial, policy, planning, and civic communities, in order for us to gain knowledge about the nature of the city's innovation network.

4 See n3, above. We have also been supported by funding from Alberta Innovates Technology Futures (AITF); the Social Sciences and Humanities Research Council of Canada; and partnerships with the National Research Council's National Institute for Nanotechnology (NINT), an internationally regarded local technology incubator; TEC Edmonton; and Northern Alberta Institute of Technology's initiatives around training for nanotechnicians and practical access to nanotechnology facilities for local fabricators, which deserves study on its own.

5 However, the "top-down" approach was recently restated as federal government policy in December 2014 (Industry Canada, 2014).

6 We reference our respondents as simply focus group participants or numbered participants when quoting an exchange in groups.

126   Rob Shields

AITF–Alberta Advanced Education and Technology. (2007). *Alberta nanotechnology strategy: Unleashing Alberta's potential*. Edmonton: Government of Alberta.

Bair, J. (2008). Analysing global economic organization: Embedded networks and global chains compared. *Economy and Society, 37*(3), 339–64. http://dx.doi.org/10.1080/03085140802172664

BCNI (Business Council on National Issues). (1993). *Building a new century economy: The Canadian challenge*. Ottawa: Business Council on National Issues.

Beaudry, C., & Schiffauerova, A. (2011). Is Canadian intellectual property leaving Canada? A study of nanotechnology patenting. *Journal of Technology Transfer, 36*(6), 665–79. http://dx.doi.org/10.1007/s10961-011-9211-1

Boltanski, L., & Thévenot, L. (2006). *On justification: Economies of worth*. Princeton Studies in Cultural Sociology. Princeton: Princeton University Press.

Bourdieu, P., & Wacquant, L. (1992). *An invitation to reflexive sociology*. Chicago: University of Chicago Press.

Brown, J.R. (2000). Privatizing the University – the new tragedy of the commons. *Science, 290*(5497), 1701–2. http://dx.doi.org/10.1126/science.290.5497.1701

Brownlee, J. (2014). Irreconcilable differences: The corporatization of Canadian universities. PhD dissertation. Department of Sociology, Carleton University, Ottawa.

Buchbinder, H., & Newson, J. (1990). Corporate-university linkages in Canada: Transforming a public institution. *Higher Education, 20*(4), 355–79. http://dx.doi.org/10.1007/BF00136218

Collier, S.J. (2009). Topologies of power: Foucault's analysis of political government beyond "governmentality." *Theory, Culture & Society, 26*(6), 78–108. http://dx.doi.org/10.1177/0263276409347694

Comaroff, J., & Comaroff, J. (2001). Millennial capitalism: First thoughts on a second coming. In J. Comaroff & J. Comaroff (Eds), *Millennial capitalism and the culture of neoliberalism* (1–56). Durham, NC: Duke University Press. http://dx.doi.org/10.1215/9780822380184-001.

Corner, A., & Pidgeon, N. (2012). Nanotechnologies and upstream public engagement: Dilemmas, debates, and prospects? In B.H. Harthorn & J.W. Mohr (Eds), *The social life of nanotechnology* (169–94). London: Routledge.

Crawford, R. (1991). Japan starts a big push toward the small scale. *Science, 254*(5036), 1304–5. http://dx.doi.org/10.1126/science.254.5036.1304

Davidson, T., Park, O., & Shields, R. (2011). *Ecologies of affect.* Waterloo, ON: Wilfrid Laurier University Press.

Davies, S.R., Selin, C., Gano, G., & Pereira, Â.G. (2013). Finding futures: A spatio-visual experiment in participatory engagement. *Leonardo, 46*(1), 76–7. http://dx.doi.org/10.1162/LEON_a_00489

Dean, M. (1996). Foucault, government, and the unfolding of authority. In A. Barry, T. Osborne, & N. Rose (Eds), *Foucault and political reason: Liberalism, neoliberalism and rationalities of government* (209–29). Chicago: University of Chicago.

Dean, M. (2009). *Governmentality: Power and rule in modern society.* London: Sage.

Degen, M. (2003). Fighting for the global catwalk: Formalizing public life in Castlefield (Manchester) and diluting public life in El Raval (Barcelona). *International Journal of Urban and Regional Research, 27*(4), 867–80. http://dx.doi.org/10.1111/j.0309-1317.2003.00488.x

Della Porta, D. & Tarrow, S. (Eds). (2005). *Transnational protest and global activism: People, passions, and power.* Lanham, MD: Rowman & Littlefield.

Drake, F. (2011). Protesting mobile phone masts: Risk, neoliberalism, and governmentality. *Science, Technology & Human Values, 36*(4), 522–48. http://dx.doi.org/10.1177/0162243910366149

Drexler, E. (1990). *Engines of creation.* New York: Anchor Books.

Elias, N., & Scotson, J.L. (1994). *The established and the outsiders – A sociological enquiry into community problems.* 2nd ed. London: Sage.

Ferguson, J., & Gupta, A. (2002). Spatializing states: Toward an ethnography of neoliberal governmentality. *American Ethnologist, 29*(4), 981–1002. http://dx.doi.org/10.1525/ae.2002.29.4.981

Foucault, M. (2007). *Security, territory, population: Lectures at the Collège de France, 1977/78.* Basingstoke, UK: Palgrave Macmillan. http://dx.doi.org/10.1057/9780230245075.

Foucault, M. (2008). *The birth of biopolitics: Lectures at the Collège de France, 1978–1979.* Basingstoke, UK: Palgrave Macmillan. http://dx.doi.org/10.1057/9780230594180.

Fuchs, S. (2001). *Against essentialism: A theory of society and culture.* Cambridge, MA: Harvard University Press. http://dx.doi.org/10.4159/9780674037410.

Gelfert, A. (2012). Nanotechnology as ideology: Towards a critical theory of "converging technologies." *Science, Technology & Society, 17*(1), 143–64. http://dx.doi.org/10.1177/097172181101700108

Ghimn, Y.-C., & Shields, R. (2015). Nanotechnology in Edmonton: An actantial diagram. Available at https://www.ualberta.ca/~rshields/f/nano/Ghimn-Shields%20Actantial%20Diagram%20of%20Nanotech%20in%20Edmonton.pdf. Accessed 14 October 2015.

Gibbons, M., Limoges, C., Nowotny, H., Schwartzman, S., Scott, P., & Trow, M. (1994). *The new production of knowledge: The dynamics of science and research in contemporary societies.* London: Sage.

Gow, B., & Sandy, A. (2007). Creating a past for the Canadian petroleum industry's technology: The drilling rig. Available at https://journals. lib.unb.ca/index.php/MCR/article/view/18089/19416. Accessed 12 September 2015.

Greenhouse, C. (Ed.). (2011). *Ethnographies of neoliberalism.* Philadelphia: University of Pennsylvania Press.

Hayek, F. (1945). The use of knowledge in society. *American Economic Review, 35*(4), 519–30.

Hayek, F. (1976). *The mirage of social justice.* London: Routledge.

Hopkins, J. (1991). West Edmonton Mall as a centre for social interaction. *Canadian Geographer / Le Géographe canadien, 35*(3), 268–79. http://dx.doi. org/10.1111/j.1541-0064.1991.tb01101.x

Hu, G.Y., Carley, S., & Tang, Li. (2012). Visualizing nanotechnology research in Canada: Evidence from publication activities, 1990–2009. *Journal of Technology Transfer, 37*(4), 550–62. http://dx.doi.org/10.1007/s10961-011-9238-3

Industry Canada. (2014). *Seizing Canada's moment: Moving forward in science, technology and innovation.* Ottawa: Government of Canada.

Irwin, A. (1995). *Citizen science: A study of people, expertise and sustainable development.* London: Routledge.

Joly, P.-B., & Kaufmann, A. (2008). Lost in translation? The need for "upstream engagement" with nanotechnology on trial. *Science as Culture, 17*(3), 225–47. http://dx.doi.org/10.1080/09505430802280727

Jones, K.E., & Irwin, A. (2012). Creating space for engagement? Lay membership in contemporary risk governance. In B.M. Hutter (Ed.), *Anticipating risks and organising risk regulation in the 21st century* (185–207). Cambridge: Cambridge University Press.

Jones, K.E., Petersen, N., Ghimn, Y.-C., . . . (2013). *Edmonton's futurescape city tours: A descriptive report.* Edmonton, AB: City Region Studies Centre.

Jones, K.E., Shields, R., & Lord, A. (2015). *City regions in prospect.* Montreal: McGill-Queen's University Press.

Joseph, J. (2012). *The social in the global: Social theory, governmentality and global politics.* Cambridge: Cambridge University Press. http://dx.doi. org/10.1017/CBO9781139149143.

JTEC(Japanese Technology Evaluation Centre) (1994). *JTEC panel report on microelectromechanical systems in Japan.* Available at http://www.wtec.org/ loyola/mems/toc.htm. Accessed 1 May 2015.

Lambert, C., Parker, A., & Neary, M. (2007). Entrepreneurialism and critical pedagogy: Reinventing the higher education curriculum. *Teaching in Higher Education*, 12(4), 525–37. http://dx.doi.org/10.1080/13562510701415672

Larner, W. (2001). Governing globalization: the New Zealand call centre attraction initiative. *Environment & Planning A*, 33(2), 297–312. http://dx.doi.org/10.1068/a33159

Larner, W., & Le Heron, R. (2002). The spaces and subjects of a globalising economy: A situated exploration of method. *Environment and Planning. D, Society & Space*, 20(6), 753–74. http://dx.doi.org/10.1068/d284t

Lave, R., Mirowski, P., & Randalls, S. (2010). Introduction: STS and neoliberal science. *Social Studies of Science*, 40(5), 659–75. http://dx.doi.org/10.1177/0306312710378549

Lorentzen, A. (2008). Knowledge networks in local and global space. *Entrepreneurship and Regional Development*, 20(6), 533–45. http://dx.doi.org/10.1080/08985620802462124

Marrtila, T. (2013). Whither governmentality research? A case study of the governmentalization of the entrepreneur in the French epistemological tradition. *Qualitative Social Research*, 14(3) n.p.

Mason, J. (2002). *Qualitative researching*. London: Sage.

Miller, P., & Rose, N. (1990). Governing economic life. *Economy and Society*, 19(1), 1–31. http://dx.doi.org/10.1080/03085149000000001

Moore, K. (2008). *Disrupting science: Social movements, American scientists, and the politics of the military, 1945–1975*. Princeton: Princeton University Press.

Moore, K., Kleinman, D.L., Hess, D., & Frickel, S. (2011). Science and neoliberal globalization: A political sociological approach. *Theory and Society*, 40(5), 505–32. http://dx.doi.org/10.1007/s11186-011-9147-3

Oels, A. (2005). Rendering climate change governable: From biopower to advanced liberal government? *Journal of Environmental Policy and Planning*, 7(3), 185–207. http://dx.doi.org/10.1080/15239080500339661

Patchett, M., & Shields, R. (Eds). (2012). *The design competition as public engagement method in strip-appeal*. Edmonton, AB: Space and Culture Publications.

Pellow, D.N. (2007). *Resisting global toxics: Transnational movements for environmental justice*. Cambridge, MA: MIT Press.

Porter, M.E. (2000). Location, competition, and economic development: Local clusters in a global economy. *Economic Development Quarterly*, 14(1), 15–34. http://dx.doi.org/10.1177/089124240001400105

Pyykkönen, M. (2007). Integrating governmentality: Administrative expectations for immigrant associations in Finland. *Alternatives*, 32(2), 197–224. http://dx.doi.org/10.1177/030437540703200203

Rose, N. (1999). *Powers of freedom: Reframing political thought.*
Cambridge: Cambridge University Press. http://dx.doi.org/10.1017/
CBO9780511488856.

Rose, N., & Miller, P. (1992). Political power beyond the state. *British Journal of
Sociology, 43*(2), 173–205. http://dx.doi.org/10.2307/591464

Schuetze, H., & Bruneau, W. (2004). Less state, more market: University
reform in Canada and abroad. *Canadian Journal of Higher Education, 34*(3),
1–12.

Shields, R. (1991). *Places on the margin: Alternative geographies of modernity.*
London: Routledge.

Shields, R. (2003). *The virtual.* London: Routledge.

Shields, R., Jones, K.E., Petersen, N., . . . (2013). *A citizens' summit:
Nanotechnology and the community. Process summary.* Edmonton, AB: City
Region Studies Centre.

Sluggett, B., Shields, R., & Jones, K. (2015). Institutional isolation (draft).
Available at http://www.ualberta.ca/~rshields/ii.pdf. Accessed 12 May
2015.

Varlander, S. (2007). The role of local strategies on a globalizing market: An
exploration of two service industry cases. *Space and Culture, 10*(4), 397–417.
http://dx.doi.org/10.1177/1206331207305830

Wolfe, D.A., & Gertler, M. (2004). Clusters from the inside and out: Local
dynamics and global linkages. *Urban Studies (Edinburgh, Scotland), 41*(5–6),
1071–93. http://dx.doi.org/10.1080/00420980410001675832

6 Neoliberal Numbers: Calculation and Hybridization in Australian and Canadian Official Statistics

COSMO HOWARD

Numbers dominate contemporary governance. Governments are expected to set and meet quantitative targets in all major policy areas. Accountability is now closely tied to performance measurement and auditing. Governmental achievements are persistently ranked against global standards and "best practices." Political fortunes and policy proposals are constantly assessed via polls, while programs are increasingly systemically evaluated using econometric techniques. Key "official statistics" such as unemployment and inflation rates are monitored by international financial markets, which can rise and fall dramatically in response to changes in the value of these indicators. As a result of this apparent hegemony of quantification, some scholars argue that *calculation* is a defining feature of neoliberal rule (Haggerty, 2001a; Rydin, 2007; Sokhi-Bulley, 2011). Calculative practices such as "numeration," "quantification," and "datafication" (Mayer-Schönberger & Cukier, 2013) appear to facilitate key reform agendas associated with neoliberal rationalities of government, including the increasing moves to calculate the worth of public goods and services in financial terms, along with the seemingly relentless push to ration government programs, make their administration more efficient, and introduce market principles into public service provision.

While compelling links can be drawn between particular instances of quantitative calculation and neoliberal "political rationalities" (Higgins & Larner, 2010; Rose, 1991), these claims confront a problem: governmental reliance on numbers clearly predates neoliberalism (Porter, 1986; Starr, 1987). The modern "science" of statistics emerged in the nineteenth century, states established large statistical bureaus in the early twentieth century, and governments came to depend extensively on numbers during the Keynesian Welfare State Era, long before the neoliberal push to retrench big government. The question this chapter asks is: how are these statistical agencies changing under the influence of neoliberal rationalities?

In keeping with the spirit of this volume, I do not assume the existence of a unified, global, and hegemonic governmental logic called neoliberalism that dominates government policy and determines individual behaviour. Rather, I use the term to draw attention to continuities across variegated and incomplete efforts in multiple jurisdictions to individualize and marketize the government of persons and institutions (Rose 1999). Consistent with this view, I adopt an analytical perspective inspired by neo-Foucauldian governmentality. Scholars working in the governmentality tradition place quantification in a larger category of calculative practices that seek to systematically measure the value of particular governmental subjects and objects (Miller & Napier, 1993; Power, 1999; Rose & Miller, 1992). They suggest that such practices are not politically neutral: these "technologies" tend to foster particular kinds of subjectivity, and to make certain governmental objectives thinkable and others unviable. Governmentality scholars are interested in the connections between calculation and neoliberalism, but they reject grand causal narratives, arguing instead that "neoliberal governance" should be seen as a complex and only partially coherent assemblage of calculative rationalities, technologies, and practices (see also Li, Lippert, Valverde, and others in this volume). Authors following this approach have investigated specific instances involving the introduction of new calculative practices such as standards, auditing, and performance measurement. As a recent wide-ranging volume on the subject concludes, studies of particular calculative practices tend to emphasize complexity, diversity, and resistance in quantification processes; this focus on plurality and disjuncture limits their capacity to provide a clear, overarching picture of the linkages between calculation and neoliberal rationalities (Higgins & Larner, 2010).

To further develop this important line of inquiry, I undertook a study of calculation inspired by recent ethnographic governmentality work on calculative practices (ibid.). However, I departed from the typical governmentality approach to case selection in that I deliberately avoided choosing an area of governance that has been recently quantified. Instead I selected a domain of governance historically dominated by quantification: *official statistics*. Because official statistics have always been calculative, I am able to ask: how are their calculative practices *changing* under the influence of neoliberal rationalities? The research combines analysis of official documents with in-depth interviews and observations of official statisticians in Australia and Canada. These jurisdictions represent important test cases for the dominant narrative around neoliberal calculative practices for two reasons: first, both countries have strongly pursued what are typically characterized as neoliberal reforms of their public services; and second, their national statistical agencies are among the most well established and highly regarded in the world in terms of technical competence and professional independence.

My argument in this chapter is that these agencies have experienced *hybridization* of calculative practices in a context where neoliberal rationalities are dominant. Following Latour (1993) and Miller, Kurunmaki, & O'Leary (2010), I define hybridization as the process through which ideas, institutions and practices that have historically been "purified" and "partitioned," are brought back into contact with one another, resulting in the blending of "apparently disparate things" (Miller et al., 2010: 22). Calculative practices such as strategic management, marketing, performance monitoring, cost accounting, and public valuation are increasingly layered over core statistical practices in official statistical agencies. Whereas existing governmentality research has emphasized that neoliberalism involves the introduction of quantitative calculations into governance fields historically dominated by qualitative judgments (Dugdale & Grealish, 2010; Miller et al., 2010), the experience of official statistics is partly the reverse: that is, statistical agencies have witnessed the intrusion of new qualitative forms of calculation and valuation into their traditionally quantitative domains.

This multiplication and diversification of calculative practices leads to *calculative hybrids*, entities that explicitly combine various organizational models and disciplinary technologies of calculation (Miller et al., 2010). The notion of hybrid is better suited to these new forms than the widely used concept of *assemblage*. Hybridization conveys the sense of a relatively deliberate effort to combine a small number of discrete practices into a coherent and stable entity, whereas assemblage suggests a complex, de-centred and unstable collection of many diverse elements. The case studies presented here reveal that statistical hybridization is a conflict-ridden process, and they show disagreement within neoliberal rationalities about the kinds of calculative practices that are required to govern effectively. This research therefore challenges much of the emerging literature that understands calculation as a technology of neoliberalism.

The chapter begins with a discussion of how calculative practices are conceptualized in the Foucauldian governmentality tradition. The following section addresses the study methodology. Subsequently, I report on both case studies, while the discussion section reflects on the implications of my findings for the broader relationship between neoliberalism and calculation.

Calculation and Governmentality

This section addresses existing scholarship on the relationship between calculation and governance. It defines calculation as the practice of applying more or less formalized techniques to determine the value of objects and subjects (Bloomfield & Hayes, 2009; Higgins & Larner, 2010; Latour, 1987, 1993). Scholars in several disciplines, including history, philosophy, and sociology, have remarked upon the rise of calculation and its close association with the emergence of the modern nation-state (Gigerenzer, Swijtink, Porter, Daston, Beatty, & Kruger, 1989; Hacking, 1990; Porter, 1986). Here I focus on the Foucauldian tradition, especially those researchers who subscribe to the governmentality approach, because they have done the most to develop conceptual linkages between calculation and governance. In this section I argue that studies of governmentality suffer from a weakness of tending to identify any instance of quantification "with neo-liberal elements as essentially neo-liberal, and to proceed as if this subsumption of the particular under a more general category provides a sufficient account of its nature or explanation of its existence" (Rose, O'Malley, & Valverde, 2006: 97–8). I suggest the proliferation of studies providing examples of "neoliberal quantification" has the unintended effect of collectively painting a portrait of strong interconnections between quantification and neoliberalism.

Early Foucauldian scholarship on calculation and governance addressed the pivotal role of statistics in the emergence of liberal government. In Foucault's seminal 1978 lecture on governmentality (Burchell, Foucault, & Senellart, 2009: see chap. 4), he argues that liberal government, that is, governing societies in accordance with the "nature of things," was made possible by statistics. As techniques of measuring populations spread in the eighteenth and nineteenth centuries, they "gradually reveal[ed] that the population possesses its own regularities:

its death rate, its incidence of disease, its regularities of accidents. Statistics also shows that the population also involves specific, aggregate effects and that these phenomena are irreducible to those of the family: major epidemics, endemic expansions, the spiral of labour and wealth. Statistics also shows that through its movements, its customs, and its activity, population has specific economic effects" (ibid.: 104).

Scholars have applied these Foucauldian insights to generate influential portraits of the historical production of statistics (Anderson, 2006; Hacking, 1990, 1999; Haggerty, 2001b; Rose, 1991; Walters, 2000). For these scholars, statistics do not simply illuminate pre-existing things such as "public opinion" or "unemployment," but they "make up" these realities (Hacking, 1999; Law & Urry, 2004). They involve the enactment of categories and the organization of events and actors into discrete sociopolitical containers. Calculation involves "conversion" of messy events into standardized categories via "coding," which identifies the "essences" of things, erases nuances, and produces a comprehensive, well-ordered "totalizing classificatory grid" (Anderson, 2006, 188). The development and implementation of "official" categories such as those used in population censuses allows for subjects and objects to be compared and united across the vast geographic and sociopolitical distances of the nation state (Latour, 1987; Rose, 1991).

More recent governmentality work explores how quantification connects with "neoliberal governance." No single definition of neoliberalism is preferred within this literature, though elements relevant to calculation include: *retrenchment*, where rights-based and redistributive state social programs are scaled back (Wacquant, 2010); *marketization*, where principles of competition, user-funding, cost recovery, and profitability are introduced into spheres of public and private life formerly governed by traditional, legal, and/or bureaucratic arrangements (Dean, 1998); *responsibilization*, where subjects inside and outside the state are enlisted in their own governance (Rose, 2000); *government at a distance*, where the state acts not in a hierarchical command fashion but through subtle indirect mechanisms of persuasion (Rose, 1999), and *individualization* and *targeting*, where actors are decreasingly addressed as members of collectives, but are treated as bearers of unique characteristics who must be governed via tailored strategies (Henman, 2007).

Governmentality scholars have addressed how these neoliberal transformations are made possible by new calculative practices both inside and outside the state. Haggerty, for instance, argues that "neo-liberal strategies of governance ... rely extensively on statistical knowledges to implement, rationalize and manage governmental programs" (2001a: 707). Henman (2007) suggests government statistical analysis has shifted throughout the twentieth century from a focus on fixed, mutually exclusive categories (e.g., employed vs. unemployed) to investigation and differentiation of individuals' personal "risk profiles" (see also Haggerty, 2001a).

At the same time as statistics are redesigned to govern citizens differently, this literature has demonstrated that new calculative practices also are introduced inside the neoliberal state. Governmentality authors focus attention on the introduction of new calculative organizational "disciplines" such as auditing, accounting, and performance measurement (Miller et al., 2010) into public sector realms traditionally dominated by professional judgment or bureaucratic proceduralism. Their ascendency inside government is strongly associated with neoliberal reform paradigms such as the new public management (NPM) (ibid.).

Some governmentality scholarship on quantification and calculation addresses the relationship between new calculative practices and existing governmental techniques (ibid.). This is particularly true of work on new calculative practices inside the state. Work on the changing role of state professions is an example where the new calculative practices are said to meld with existing professional practices to produce "calculative hybrids." Miller et al. (2010) give the example of the implementation of NPM reforms in health and social care, where caring professionals are increasingly subjected to quantitative reviews of their performance and also expected to employ quantitative "evidence" in their treatment of service users and management of subordinates. They are increasingly required to draw on techniques informed by knowledge from multiple disciplines: economics, accounting, auditing, management, and human resources. These new calculative practices do not replace established professional, qualitative forms of calculation, but instead operate as a series of problematizations and questions (cf. Collier, 2009, 2011) that practitioners should consider in carrying out their professional work: Can I improve the efficiency of my core work practices? Am I being sufficiently responsive to stakeholders? Am I adding value through my work and, if so, how much? Am I exploiting my staff's full potential? Am I contributing to the financial sustainability of my working environment? The superimposition of these problematizing calculative practices leads to a process of "disciplining the disciplines" where the core professional/disciplinary work of expert administrators comes to be scrutinized via other disciplinary practices (Dugdale & Grealish, 2010).

This governmentality scholarship is valuable because it brings Foucault's insights about the early modern governmental implications of ostensibly neutral statistical technologies and practices into a new era. However, there are several weaknesses. While individual works eschew grand narratives of ideological or technological determinism, together they have several (unintentional) and potentially misleading cumulative effects on how we understand the connection between quantification and contemporary governance (cf. Brady, 2014). First, the focus on quantification of domains previously governed without numbers suggests a general intensification of quantification. Second, this focus is exacerbated by the concentration on new calculative practices. The result is a tendency to imply a necessary relationship between an intensification of quantification and the dominance of neoliberal rationalities.

Methodology

To study how practices of calculation are changing in a context where neoliberal political rationalities are dominant I engage in what might be broadly referred to as an ethnographic approach to an analytics of governmentality (Brady, 2011, 2014). In contrast to traditional interpretive research, which foregrounds actors' creative agency, downplays the role of rationalities and problematizations in structuring experiences, and commonly contrasts "what really happens" with the fictions in governments' plans, this approach does not privilege individuals' narratives over other sources but instead treats them equally as a source of insight into problematizations, rationalities, and technologies for governing (see also both Dorow and Larner in this volume).

I used a combination of textual analysis, in-depth interviews, and observation. In line with the governmentality tradition, my starting point was a systematic discursive analysis of documents, including annual reports, official administrative histories, and major publications addressing strategic changes in official statistics. These texts conveyed key historical shifts, the official organizational lexicon, and the development and implementation of initiatives in areas of statistical methodology, organizational management, and user engagement. However, beyond summary treatments of "implementation challenges" they provided little discussion of problems and tensions surrounding the implementation of various calculative practices in the agencies.

To delve further into practices and experiences of calculative hybridity I interviewed current and former senior statistical officials (fourteen in Australia between 2005 and 2014[1] and twelve in Canada between 2010 and 2012). Prospective participants at the statistical agencies were invited by approaching their chief executives and asking to interview senior staff involved with statistical governance. Interviewees included staff responsible for executive leadership, budgeting, human resources, communications, methodology, and major program areas such as the census, social surveys, and economic indicators. Participants were asked to describe their various calculative practices, including statistical methodology, strategic management, accounting, and client relations, and also to explain how these procedures were integrated within the agency.

Visits to statistical agencies included several ethnographic elements of participant observation. Some interviews began (unexpectedly) with slide presentations, where I was positioned as a recipient of factual summaries of organizational practice and performance, but which also allowed me to observe organizational performance. Some interviews took place in a group format, where staff would sit around a small meeting table in an office, or at a large boardroom table, and take turns explaining their roles and answering questions. This information helped to reveal intra-agency relationships and dynamics, including the interactions between staff responsible for different calculative practices. Such ethnographic elements added greater depth to the documentary and interview material and exposed problems of governance that otherwise would not have been visible to me.

Two Case Studies of Calculation in Contemporary Official Statistics

To determine how calculative practices in official statistical agencies are changing under neoliberalism we need to establish how these agencies historically engaged in calculation. There is insufficient space here to discuss the history of the Australian Bureau of Statistics (ABS) and Statistics Canada (StatCan) (for comprehensive histories see ABS, 2005; Worton, 1998; Worton & Platek, 1995). For current purposes it is sufficient to summarize two important historical calculative traditions shared by both the ABS and StatCan (along with many other national statistical agencies; see Starr, 1987). First, both agencies were created in the early twentieth century[2] to capitalize on the emerging science of statistics by building a critical mass of experts in the new techniques of social and economic measurement and enumeration. Second, the agencies were intended to be independent of the government of the day, meaning that cabinets and individual ministers were expected to refrain from instructing the agencies on how to spend their budgets. They were not servants of other departments and agencies or other levels of government, but were neutral, disinterested, steady producers of statistics (ABS, 2005; Worton & Platek, 1995). Both the ABS and StatCan have consistently maintained that *they alone decide* what they measure and how it is measured (ibid.). In essence, these agencies were created to focus exclusively on statistical practices, and they incorporated a mixture of legal protections and de facto conventions to prevent the encroachment of non-statistical considerations into their work.

How have these agencies' calculative practices changed with the rise of neoliberal rationalities? The next two sections separately address Australian and Canadian experiences since 1990.

Australia: Statistical Expertise Confronts the New Public Management

In this section I argue that in recent decades the Australian Bureau of Statistics has confronted governance pressures to introduce new calculative practices of cost accounting, public valuation, auditing, and risk management, which have produced hybridization, along with conflicts, disruptions, and resistances within the agency.

While the ABS enjoys a high degree of legislated independence from the government of the day, it has always been potentially vulnerable to external pressure via its budget, which is reviewed by the Department of Finance and must be approved by cabinet and Parliament. According to one interviewee, the ABS did not historically face budget pressure because of its location within the powerful Commonwealth Treasury (the ministry responsible for the federal budget), which insulated it from cabinet interference (former ABS official).

However, in the early 1990s this situation changed. Like many Anglo-American jurisdictions, Australia went through a period of new public management reforms, driven with particular enthusiasm in the Australian case by the Department of Finance (Schwartz, 1994). Changes included the introduction of rolling annual budget cuts for federal bodies (called "efficiency dividends"), new pressure to introduce better internal accounting for costs, and encouragement to increase agency reliance on user funding (Wanna, Forster, & Kelly, 2000). A former ABS senior executive recalled how Finance imposed this requirement to

charge end users a price on the agency and how it clashed with the attitudes of ABS staff, who felt they were providing a free public good paid
for by the general budget appropriation: "What had happened was ...
the Department of Finance just did an arbitrary cut on the ABS budget
and said, 'You raise charges for some of your outputs.' And [that] was
massively unpopular because you were charging for things that were
basically free before and which were seen to be funded by appropriation. So that was very unpopular" (former ABS official). Nevertheless,
this intervention did change staff attitudes and behaviours: "Charging
was introduced in order to have an impact on clients' behaviour. But
it also had an impact on staff, who were trying to push cost recovery.
It triggered certain behaviours, a dynamic of operating along business
lines" (ABS official).

As a result, under the leadership of Australian Statistician Ian Castles the ABS began a program of "cost recovery" statistics. These were
figures produced by the ABS for specific clients (almost always other
federal agencies or Australian state governments) who would fund the
work. The cost recovery program encouraged a series of qualitative and
quantitative calculations to be made by ABS staff. The most important
was a calculation about the alignment between proposed externally
funded work and the agency's work program and "mandate" (former ABS official). Closely related was a public value calculation that
required the agency officials to determine whether the proposed contractual work would be in the public interest: "We made sure that we
were not doing things that didn't fit within, you know, the Bureau's
mandate. That it was public information of interest. Anything that we
collected had to be put out in the public domain in the same way that
we would for any other publication. So there was nothing secret, or
market research or anything like that about it. As I said it was mainly,
I think it was all for public customers. It probably didn't have to be as
a matter of policy, but it had to be essentially a public good" (former
ABS official).

Finally, staff had to calculate costs and determine a reasonable price
for the work, which was challenging, given that production functions
are spread across multiple divisions in the agency, and in some areas of
statistics there is no market price to benchmark against (ABS official).
This NPM-inspired cost-recovery program encouraged the agency to
adopt the position of a calculative hybrid: it would remain a bureaucratic base-funded organization dedicated to the public interest but
also would embrace a market-driven logic. It would combine pure

statistical science with a test of utility based on client demand, and it would need to develop new calculative practices and meld multiple disciplinary knowledges, including revenue forecasting, cost accounting, and statistical methods.

While some in the agency celebrated this new calculative hybridity, others were hostile. One client-funded survey in particular – the Women's Safety Survey (WSS) – exposed the diverging world views of the statistical purists versus supporters of the new hybridity. This survey was administered by the ABS but funded by the Office of the Status of Women within the Department of the Prime Minister and Cabinet. Its development in the mid-1990s was motivated by recognition of the paucity of Australian data on domestic violence relative to comparable jurisdictions (Coochey, 1995). Yet, before the survey was even implemented, criticism emerged from within the agency (ABS, 2005: 133; Coochey, 1995). According to one former ABS official, "[The WSS] caused a lot of controversy. We had a couple of staff in the place who went public about that … it got quite heated." The source of the controversy was apparently the fact that only women would be surveyed: "there was a lot of angst from [ABS staff] who said you should do a survey of both sexes. And my response was, well, that would be very nice if there was an Office of the Status of Men who wanted to fund that. I'd be quite happy to do that … The safety of men wasn't an issue. The issue … was the safety of women and it was perfectly reasonable to do that … but if you didn't have money, well, that was tough" (former ABS official).

The opposed staff members went to the *Canberra Times* newspaper with their concerns. One disparagingly asserted that the ABS "has only one aim in life, i.e., to turn a profit by any means possible" (Coochey, 1995). The Office of the Status of Women (OSW) funding also led to concerns within the ABS that "*she* who pays the piper should be allowed to call the tune," and this would "discredit other work done by the ABS, which is supposed to be the national official independent statistical agency" (ibid.; emphasis added). Despite the ABS defending the survey results as valid, the controversy "set the climate for clamping down on this sort of work" (former ABS official). The WSS ran for only two years before being cancelled. ABS interviewees suggested that the agency now does very little contractual work of this sort. Thus, the Women's Safety Survey episode illustrated the extent of concern within the agency about hybridizing calculative practices and the desire to protect the purity of the ABS statistical work from being tainted by revenue considerations and "ideological" agendas.

More recent developments suggest there is an ongoing disagreement about the appropriate role for these new market-driven calculative practices and rationalities within the agency. A prime example is the ABS's role in longitudinal data. In the late 1990s the ABS was approached to oversee a program of federal government longitudinal data sets in social policy. The responsible federal department was keen to fund longitudinal data sets so that researchers and policy makers could investigate trends in employment, income support receipt, children's well-being, and Indigenous life outcomes (Wooden & Watson, 2007). The Australian statistician at the time, Bill McLennan, rejected the longitudinal work on the grounds it created unacceptable risks for the ABS's reputation: longitudinal data[3] is subject to greater statistical error, leading to potentially embarrassing mistakes in statistical releases, and it can compromise respondent confidentiality because the same individuals are being surveyed over time and are therefore easier to identify in the data. However, according to an interviewee, McLennan's replacement, Dennis Trewin, "rued the fact" that McLennan had passed up the longitudinal work, because he realized it had great potential as a growth area and felt it was possible to manage the risks with some sensible precautions (Australian non-government statistician). According to this interviewee, when new longitudinal data sets were announced, Trewin offered the ABS's services for data collection only, leaving another organization to process the data. These statistics were notionally distanced from the ABS, so the agency was not directly exposed to the risk of errors and confidentiality problems. As a result the ABS took on contract data collection work for several national longitudinal surveys. Despite this compromise based on a calculative logic of risk management, Trewin's replacement, Brian Pink, felt it too risky for the bureau's reputation to be involved in even an indirect capacity, and he put a stop to new longitudinal data collection contracts.

Over the same period, during which the ABS has largely rejected external cost recovery, new calculative practices have also appeared in the day-to-day running of ABS operations. An important example is the introduction of "internal cost recovery" in relation to information technology services, whereby divisions are charged when they use IT support and infrastructure development. One staff member expressed frustration at the new requirement: "It's a pain. It's a convoluted process. In a sense, it's 'funny money'" (ABS official). Nevertheless, this official acknowledged that "it adds some rigours to the process." Interviewees indicated that internal cost recovery forced staff to engage in new calculative practices: it "created very useful dynamics" by enabling "a more structured consideration" of the resources required for particular products, thereby allowing the agency to "set [internal] prices for infrastructure services." Ultimately, the move introduced a new discipline and way of thinking into the organization: "Cost recovery has been a good thing. It forces people to look at real costs. So it can be hard to come to grips with" (ABS official). ABS interviewees argued that, despite some initial unpopularity, internal cost recovery is now regarded as an important tool of coordination and planning. It is part of a broader effort to shift the culture in the ABS from being exclusively focused on methodological concerns to also thinking about strategic management issues and priorities. As another interviewee put it: "Our top people are extremely capable in terms of methodology. That is a great asset but also a liability at times. They want to get down into the details but sometimes it bogs down and our big challenge is to have a discussion about the strategic issues" (ABS official).

Thus, in recent years the ABS has experienced significant pressures to hybridize its practices by introducing a greater role for market rationalities and relationships as well as embarking on new calculative practices such as cost recovery, demand-driven pricing, and public valuation. The push for calculative hybridization has generated resistance. While we cannot say such practices have been definitively rejected, they nevertheless produce internal disagreements leading to inconsistent, contested hybridization.

Canada: Calculative Contradictions

In contrast to the situation in Australia, Canadian official statistics did not experience substantial shifts in the 1990s, for two reasons. First, NPM has not had the same kind of impact in Canada as in Australia. Whereas Australian agencies were directly pressured to adopt new operating models such as user funding and internal cost recovery, NPM at the federal level in Canada mostly took the form of funding cuts in the mid-1990s (Aucoin, 2002). Second, Canada employed the same chief statistician (Ivan Fellegi) from 1985 to 2008, which, according to interviewees, led to significant continuity of operations and forestalled reforms that might otherwise have occurred. In this section I focus on a more recent event that dramatically overturned this long period of continuity: the 2010 cancellation of StatCan's flagship mandatory long-form census (MLFC). I suggest that the episode has forced the agency to embrace multiple calculative practices and exposed the political agendas, discourses, and conflicts underlying the census.

Canada carries out a census of the population every five years. In accordance with statistical legislation, the Canadian cabinet was asked in January 2010 to approve the 2011 Population Census, including the "short-form" component (a short questionnaire mailed to all households) and the "long-form" component (a much longer demographic survey mailed to only one in five households to save costs). Approval in the past had been a routine formality, but on this occasion the census was held up in cabinet and the decision deferred. According to an interviewee, some ministers in the Conservative government led by Stephen Harper objected to the mandatory nature of the census and the threat of fines and jail time for non-compliance: "There was flak in cabinet ... from radical and to some degree paranoid American views about the census. You don't have to read a whole lot to see the sort of conspiratorial ideas. Government's going to take our guns away ... Libertarianism combined with the conspiracy views" (Canadian official).

In June Industry Minister Tony Clement announced that the mandatory long-form census would be scrapped and replaced by a voluntary "National Household Survey," which essentially would use the same questions as the MLFC. The short-form census would remain mandatory. The decision was justified using the same arguments that had emerged in cabinet about the unfairness of coercing Canadians to reveal intimate details of their personal arrangements in the original long form. In addition, the industry minister claimed that he had decided to scrap the mandatory long-form census following advice from Statistics Canada that a voluntary survey would produce data as representative as a compulsory one. This claim put StatCan's credibility as a statistical agency under threat. A member of the National Statistics Council, an independent advisory body, put the issue starkly to Chief Statistician and head of StatCan Munir Sheikh in a confidential memo (obtained under access to information): "Your minister is attempting to use you as cover for partisan purposes" and "you have no choice but to resign" (Postmedia News, 2011). Sheikh resigned on 21 June and explained his position to a House of Commons select committee on 10 July: "When doubt began to be expressed about the nature of the advice we gave, which to any statistician would come across as not the work of a statistician, I came to the conclusion that I cannot be the head of an agency whose reputation has suffered" (House of Commons, 2010: 1044).

Why did a government usually characterized as neoliberal (Ramp & Harrison, 2012) attack this key calculative practice? Most observers do not believe that the stated rationale – concern about coercion and privacy – fully explains the motivations and strategies involved. Instead, commentators have generally agreed the cancellation was a political tactic to both appeal to a conservative voting base and undermine data used to support social programs. Scrapping the MLFC furthered these objectives in several ways. It constructed an image of a government taking on and disarming a coercive federal bureaucracy (ibid.; Wells, 2013). Ramp & Harrison (2012) suggest that the Harper government was also appealing to a "folk epistemology" popular on the political Right: the view that there is nothing intrinsically superior about facts produced in a technically complex Ottawa bureaucracy compared with the authentic knowledge of an ordinary person. These observers argue that the government also sought to appeal to supporters by portraying the census as a political instrument used by promoters of social welfare, First Nations, and minority rights; the census made the plight of disadvantaged groups visible, thereby justifying state spending on social programs (Wells, 2013). Undermining the census was thus seen as a strategy for concealing existing disadvantage as well as the deleterious social impacts of future neoliberal reforms. For this reason, the criticisms by some advocacy groups that emerged during the controversy played into the government's hands, as this Canadian official observed: "The last thing we need is someone out there complaining about how their research into marginalized and oppressed immigrant women will be harmed by [cancelling the MLFC]. You know, the Prime Minister's Office will ... run laps around the building they'll be so happy."

Thus, eliminating the MLFC has been portrayed as motivated by both short-term electoral tactics of appealing to a subset of voters, as well as a longer-term strategy designed to undermine redistributive social policy by dismantling part of the infrastructure of the Canadian welfare state. In this view undermining these calculative practices was entirely consistent with a neoliberal retreat of the state from private life and the retrenchment of social programs.

Yet the story is not so simple. Throughout the controversy actors who had strongly promoted neoliberal rationalities, including welfare retrenchment, market discipline, and governing at a distance, were among the most active critics of the decision to eliminate the MLFC. According to one interviewee, early in the process both the minister of finance and the minister of industry raised serious concerns about the economic implications of the change: "Letters had gone from the minister of finance and the minister of industry to the prime minister expressing their concern about this ... It caused a fair stir at a strato-spheric level of the bureaucracy, when you get ministers writing or communicating in some formal fashion to the prime minister saying this has serious issues related to it" (Canadian official).

According to this interviewee, the ministers warned the PM of serious negative economic and administrative consequences should the change proceed. The removal of the mandatory requirement would compro-mise data quality and in turn risk the trust of international markets in Canada's economic performance data. Also, the ministers pointed to the heavy dependence of private sector decision makers on census data when making investment and production decisions. Here we see two ministers making a (private) case for the census on economic grounds, as a key instrument of a growth-focused policy agenda. Other actors traditionally aligned with a small government and free-market position also objected. Charlie Angus, an MP for the left-leaning New Democratic Party, argued before the House of Commons that Industry Minister Clement had dis-missed the concerns of "Quite a divergent group, including ... key bank-ers, municipal planners, provincial thinkers, all the top bank economists ... [They] tried to meet with you in order to address this issue and try to find solutions, and you blew them off" (House of Commons, 2010). This led the minister to observe: "Hon. Tony Clement: I think I've gone through the looking glass, Mr Chair, where the NDP is demanding that I meet with bankers. Voices: Oh, oh!" (House of Commons, 2010).

These protests reveal diverse arguments about the political and gov-ernmental implications of eliminating the MLFC. While for some the MLFC decision would facilitate the retrenchment of social programs, for others the decision had the potential to harm economic growth and the quality of market decision making. Of course, these two arguments are not necessarily mutually exclusive, but they highlight divergent perspectives about the relationship between a specific calculative prac-tice and broader neoliberal political rationalities. They show how the very same statistical technology (in this case the long-form census) can be simultaneously defended as necessary for the smooth functioning of the market and challenged as an impediment to the liberation of mar-kets from the shackles of regulation, welfarism, and the demands of special interests.

As a result of the MLFC episode and a subsequent series of (government-wide) budget cuts, StatCan felt pressure to engage in new calculative practices, including quality review, demand fore-casting, and strategic management. In the wake of the MLFC contro-versy, there was considerable concern that the replacement National Household Survey would produce results much less trustworthy than the previous MLFCs. To determine the difference in quality, Statistics Canada initiated a review of the 2006 MLFC data to set a baseline for comparison with the voluntary survey. This procedure revealed some surprising, previously overlooked weaknesses in the data: "In the process of dealing with 2011 we made some discover-ies about 2006 that ... didn't make us very comfortable. The sam-pling variability in small areas and for small groups was so high that you really couldn't use the data ... So we actually discovered that the quality of some of the data from 2006 was vastly worse than we ever thought it was and we published it with a warning" (StatCan official).

While StatCan has always conducted quality checking, this new environment of a voluntary survey makes it more imperative to have a system in place and not merely to produce the statistics but to stand back and make calculative judgments about their quality:

> The bottom line is I'm not saying that we're not going to encounter prob-lems with the data. But we just encountered some problems with the [2006 MLFC] data ... All I'm saying is that you cannot make that judgment until you have the data in hand and you've analysed it and that's what we're doing now. So we will conduct our analysis, we will do data certification, we will conclude whether there are issues with the data. If we find in the data issues that make the data invalid, we will choose not to publish it and we will issue a statement that this data will not be available because we have encountered problems of this type and we don't believe it's useable. (StatCan official)

Thus, while the calculative practices associated with statistical quality checking in themselves are not new at StatCan, the MLFC episode has encouraged a heightened reflexivity with a more systematic and self-critical process of assessing the quality of existing and future statistical outputs and reporting publically about data problems.

In the context of budget austerity StatCan has also applied organi-zational design and management practices to enhance production effi-ciency. One interviewee argued that the long, stable, and successful tenure of Ivan Fellegi removed pressure from senior management to develop a systematic or "corporate" approach to allocating resources throughout the agency and encouraged several inefficiencies in task allocation. This official stated that a more strategic approach was made necessary by the funding cuts: "We need to ... modernize or industrialize, I guess modernize in this case our whole statistical production process. We have a major initiative underway. We call it Corporate Business Architecture to streamline the way we do business, to make our systems more efficient, more robust, more responsive" (StatCan official).

As part of these reforms senior management went through a process of asking critical questions about the existing statistical production arrangements: "Why do we have collection activity being done in twenty divisions when in fact if we centralize it we could professionalize it and make it more efficient? Why do we have ... like sixty divisions developing tools, when we only need one?" (StatCan official). Such systematic questioning of established statistical production practices generated fear and resistance in some parts of StatCan, but senior management regarded it as essential:

> It's frightening to some parts of the organization ... but resistance is not an option. Resistance is futile. And the alternative is ... equally frightening. Trying to stay the way we worked five years ago, [pause] you could try to do that but the world is changing. And it's not that the past didn't serve us extremely well; it did, you know. It's just there are enough different pressures coming at us from different angles that we need to rethink how we deliver our business lines. And I think that is a healthy conversation that organizations should have on a regular basis. We're having it now. Part of it is being driven by ... a series of budget cuts, and the cumulative effect is really forcing us to look for efficiencies within the organization – but efficiencies we can invest back into the organization and it's to maintain our relevance. It's absolutely ... to modernize the infrastructure that is supporting us. We cannot have an infrastructure that has fifty different processing environments. (StatCan official)

Thus, StatCan's recent experiences have encouraged a layering of calculative practices. Existing statistical production work is increasingly subject to a regime of quality auditing, as well as managerial scrutiny to determine if work can be reorganized to make it more efficient and responsive. This layering appears to be threatening to some parts of the organization.

Discussion: Calculating the Calculators

This chapter has explored how governmental practices of calculation have changed during periods in which neoliberal political rationalities have been dominant. We saw that multiple forms of calculative practice have been taken up inside statistical agencies in recent years. In this section I reflect on what these empirical findings say about the relationship between neoliberal rationalities, calculation, and quantification.

The findings support the contention that neoliberal rationalities involve an intensification of calculation, but the research also complicates and contradicts several prevailing observations about the relationship between calculative practices and neoliberal political rationalities.

First, the intensification of calculation does not necessarily mean greater quantification. The existing governmentality literature highlights the ingress of quantitative disciplines and techniques into realms previously governed by local, qualitative, professional judgments. However, by studying cases that were historically dominated by quantification, we revealed a different picture. In official statistics the intensification of calculation has involved a shift from purely mathematical practices to a hybrid mixture of statistical, financial, managerial, commercial, and strategic calculations. Thus, while the dominance of neoliberal rationalities has been correlated with increased quantification in many areas of state activity, it is also associated with the introduction of new qualitative calculative practices in areas traditionally dominated by numbers. While neoliberal rationalities are not necessarily correlated with increased quantification, there are nevertheless two clear trends concerning calculation: (1) there is increasing hybridization of practices drawn from historically distinct calculative disciplines and rationalities; and (2) actors are increasingly required to engage in self-calculation, meaning public servants must reflexively calculate the value of their own calculative efforts.

Second, my study does not support the argument that the dominance of neoliberal rationalities has involved replacing broad, fuzzy calculations of public interest with narrower financial measures of value. Instead I showed how the introduction of market mechanisms into statistical production encouraged public officials to more directly and explicitly think about the public interest in order to uphold their public interest mandate. The shift to user-funding of agency work meant staff were routinely confronted with new project proposals and were required to regularly assess proposals to determine their public interest value (in addition to the revenue opportunities). This outcome may have been driven in part by the contemporary "public value" movement, which seeks to counteract aspects of new public management by encouraging public servants to pursue non-market benefits as their most important objectives (Moore, 1995, 2013). Yet this study shows that actors do not necessarily conceive public value as competing with market value, but often look for opportunities to align the two forms of calculation. Thus, while the ascendance of neoliberal rationalities is correlated with intensified calculation, it has not necessarily resulted in the extension of privatized measures of worth at the expense of public valuation.

Finally, my study suggests that connections between specific calculative practices and larger neoliberal political rationalities are more complex and problematic than is typically acknowledged. While governmentality scholars point to calculative hybridization as a key process through which "micro-practices" are linked to larger neoliberal governance agendas, I found hybridization was continuously challenged and resisted. In part this reflects administrative resistance to the neoliberal/NPM push for efficiency and retrenchment. Yet it is not merely resistance to neoliberalism that creates tensions within calculative hybrids. I suggest that conflicts at the level of calculative practice are also the product of divergent neoliberal rationalities. For example, the Canadian census case exposed two competing neoliberal rationalities. The first treated the census as a tool of good governance, defended by bankers and economists as essential for "sound economic management," that is, the maintenance of economic growth in a free market economy and provision of accurate data to indirectly steer and support corporate decision making. The second interpreted the census as a tool of legitimation for interventionist and redistributive social policies. In the second rationality, the census had to be dismantled to undermine political support for social programs that allegedly drain resources and hamper economic reform and growth. In other words, the census as a technology of governance was simultaneously interpreted and constructed as being consistent with and in tension with neoliberal rationalities. Ultimately, a governmentality approach provides insights into the connections between neoliberalism and technologies of calculation, and this study illustrates the insights that are possible when ethnographic methods are used to illustrate historical change and ongoing contestation that would otherwise not be visible.

NOTES

1 One of the Australian interviewees was an employee of a partially
 publically funded statistical body working outside of government.
2 The Australian Commonwealth Bureau of Census of Statistics was created
 in 1905 and renamed the Australian Bureau of Statistics in 1975. The
 Dominion Bureau of Statistics was created in Canada in 1918 and renamed
 Statistics Canada in 1971.
3 I regard plural usage of "data" as archaic and treat the term as singular.

REFERENCES

ABS. (2005). *Informing a nation: The evolution of the Australian Bureau of
 Statistics*. Canberra: Australian Bureau of Statistics.
Anderson, B. (2006). *Imagined communities: Reflections on the origin and spread of
 nationalism*. London: Verso Books.
Aucoin, P. (2002). Beyond the "new" in public management reform in Canada:
 Catching the next wave? In C. Dunn (Ed.), *The handbook of Canadian public
 administration* (37–52). Don Mills, ON: Oxford University Press.
Bloomfield, B.P., & Hayes, N. (2009). Power and organizational transformation
 through technology: Hybrids of electronic government. *Organization Studies*,
 30(5), 461–87. http://dx.doi.org/10.1177/0170840609104394
Brady, M. (2011). Researching governmentalities through ethnography: The
 case of Australian welfare reforms and programs for single parents. *Critical
 Policy Studies, 5*(3), 264–82. http://dx.doi.org/10.1080/19460171.2011.606300
Brady, M. (2014). Ethnographies of neoliberal governmentalities: From the
 neoliberal apparatus to neoliberalism and governmental assemblages.
 Foucault Studies, 18(1), 11–33.
Burchell, G., Foucault, M., & Senellart, M. (2009). *Security, territory, population:
 Lectures at the Collège de France, 1977–1978*. Basingstoke, UK: Palgrave Macmillan.
Collier, S.J. (2009). Topologies of power: Foucault's analysis of political
 government beyond "governmentality." *Theory, Culture & Society, 26*(6),
 78–108. http://dx.doi.org/10.1177/0263276409347694
Collier, S.J. (2011). *Post-Soviet social: Neoliberalism, social modernity,
 biopolitics*. Princeton: Princeton University Press. http://dx.doi.
 org/10.1515/9781400840427
Coochey, J. (1995, 9 April). OSW's one-eyed study of violence. *Canberra Times*, 9.
Dean, M. (1998). Administering asceticism: Reworking the ethical life of the
 unemployed citizen. In M. Dean & B. Hindess (Eds), *Governing Australia:*

Studies in contemporary rationalities of government (87–107). Cambridge: Cambridge University Press.

Dugdale, A., & Grealish, L. (2010). New modes of governance and the standardization of nursing competencies: An Australian case study. In V. Higgins & W. Larner (Eds), *Calculating the social: Standards and the reconfiguration of governing* (94–111). Basingstoke, UK: Palgrave Macmillan.

Gigerenzer, G., Swijtink, Z., Porter, T., Daston, L., Beatty, J., & Kruger, L. (1989). *The empire of chance: How probability changed science and everyday life*. Cambridge: Cambridge University Press. http://dx.doi.org/10.1017/CBO9780511720482

Hacking, I. (1990). *The taming of chance*. Cambridge: Cambridge University Press. http://dx.doi.org/10.1017/CBO9780511819766

Hacking, I. (1999). Making up people. In T.C. Heller, M. Sosna, & D.E. Wellbery (Eds), *Reconstructing individualism* (222–36). Stanford: Stanford University Press.

Haggerty, K.D. (2001a). Negotiated measures: The institutional micropolitics of official criminal justice statistics. *Studies in History and Philosophy of Science Part A, 32*(4), 705–22. http://dx.doi.org/10.1016/S0039-3681(01)00018-8

Haggerty, K.D. (2001b). *Making crime count*. Toronto: University of Toronto Press.

Henman, P. (2007). Governing individuality. In C. Howard (Ed.), *Contested individualization: Debates about contemporary personhood* (171–85). New York: Palgrave Macmillan.

Higgins, V., & Larner, W. (Eds). (2010). *Calculating the social: Standards and the reconfiguration of governing*. Basingstoke, UK: Palgrave Macmillan. http://dx.doi.org/10.1057/9780230289673

House of Commons. Canada. (2010). Bill C-568. An act to amend the Statistics Act (mandatory long-form census). Standing Committee on Industry, Science and Technology. Ottawa: Parliament of Canada.

Latour, B. (1987). *Science in action: How to follow scientists and engineers through society*. Cambridge, MA: Harvard University Press.

Latour, B. (1993). *We have never been modern*. Cambridge, MA: Harvard University Press.

Law, J., & Urry, J. (2004). Enacting the social. *Economy and Society, 33*(3), 390–410. http://dx.doi.org/10.1080/0308514042000225716

Mayer-Schönberger, V., & Cukier, K. (2013). *Big data: A revolution that will transform how we live, work, and think*. Boston: Houghton Mifflin Harcourt.

Miller, P., Kurunmaki, L., & O'Leary, T. (2010). Calculating hybrids. In V. Higgins & W. Larner (Eds), *Calculating the social: Standards and the reconfiguration of governing* (21–37). Basingstoke, UK: Palgrave Macmillan.

Miller, P., & Napier, C. (1993). Genealogies of calculation. *Accounting, Organizations and Society, 18*(7–8), 631–47. http://dx.doi.org/10.1016/0361-3682(93)90047-A

Moore, M.H. (1995). *Creating public value: Strategic management in government.* Cambridge, MA: Harvard University Press.

Moore, M.H. (2013). *Recognizing public value.* Cambridge, MA: Harvard University Press.

Porter, T.M. (1986). *The rise of statistical thinking, 1820–1900.* Princeton: Princeton University Press.

Postmedia News (2011). StatsCan panel tried to fight decision to kill long-form census: Documents. Available at http://ipolitics.ca/2011/03/02/statscan-panel-tried-to-fight-decision-to-kill-long-form-census-documents/. Accessed 2 March 2015.

Power, M. (1999). *The audit society: Rituals of verification.* Oxford: Oxford University Press. http://dx.doi.org/10.1093/acprof:oso/9780198296034.001.0001

Ramp, W., & Harrison, T.W. (2012). Libertarian populism, neoliberal rationality, and the mandatory long-form census: Implications for sociology. *Canadian Journal of Sociology, 37*(3), 273–94.

Rose, N. (1991). Governing by numbers: Figuring out democracy. *Accounting, Organizations and Society, 16*(7), 673–92. http://dx.doi.org/10.1016/0361-3682(91)90019-B

Rose, N. (1999). *Powers of freedom: Reframing political thought.* Cambridge: Cambridge University Press. http://dx.doi.org/10.1017/CBO9780511488856

Rose, N. (2000). Government and control. *British Journal of Criminology, 40*(2), 321–39. http://dx.doi.org/10.1093/bjc/40.2.321

Rose, N., & Miller, P. (1992). Political power beyond the state: Problematics of government. *British Journal of Sociology, 43*(2), 173–205. http://dx.doi.org/10.2307/591464

Rose, N., O'Malley, P., & Valverde, M. (2006). Governmentality. *Annual Review of Law and Social Science, 2*(1), 83–104. http://dx.doi.org/10.1146/annurev.lawsocsci.2.081805.105900

Rydin, Y. (2007). Indicators as a governmental technology? The lessons of community-based sustainability indicator projects. *Environment and Planning. D, Society & Space, 25*(4), 610–24. http://dx.doi.org/10.1068/d72j

Schwartz, H.M. (1994). Public choice theory and public choices: Bureaucrats and state reorganization in Australia, Denmark, New Zealand, and Sweden in the 1980s. *Administration & Society, 26*(1), 48–77. http://dx.doi.org/10.1177/009539979402600104

Sokhi-Bulley, B. (2011). Governing (through) rights: Statistics as technologies of governmentality. *Social & Legal Studies, 20*(2), 139–55. http://dx.doi.org/10.1177/0964663910391520

Starr, P. (1987). The sociology of official statistics. In W. Alonso & P. Starr (Eds), *The politics of numbers* (7–57). New York: Russell Sage Foundation.

Wacquant, L. (2010). Crafting the neoliberal state: Workfare, prisonfare, and social insecurity. *Sociological Forum, 25*(2), 197–220. http://dx.doi.org/10.1111/j.1573-7861.2010.01173.x

Walters, W. (2000). *Unemployment and government: Genealogies of the social.* Cambridge: Cambridge University Press. http://dx.doi.org/10.1017/CBO9780511557798

Wanna, J., Forster, J., & Kelly, J. (2000). *Managing public expenditure in Australia.* St Leonards, AUS: Allen & Unwin.

Wells, P. (2013). *The longer I'm prime minister: Stephen Harper and Canada, 2006–.* Toronto: Random House Canada.

Wooden, M., & Watson, N. (2007). The HILDA Survey and its contribution to economic and social research (so far). *Economic Record, 83*(261), 208–31. http://dx.doi.org/10.1111/j.1475-4932.2007.00395.x

Worton, D.A. (1998). *Dominion Bureau of Statistics: A history of Canada's central statistical office and its antecedents, 1841–1972.* Montreal: McGill-Queen's University Press.

Worton, D.A., & Platek, R. (1995). *A history of business surveys at Statistics Canada: From the era of the gifted amateur to that of scientific methodology.* Ottawa: Statistics Canada.

7 Governing through Land: Neoliberal Governmentalities in the British Columbia Treaty Process

AKIN AKINWUMI AND NICHOLAS BLOMLEY

Neoliberalism is everywhere, yet, at the same time, it appears to be nowhere. As a concept, neoliberalism is highly contested: it is said to be nebulous, neither unified nor uniform, slippery, promiscuous, and even illusory (see Barnett, 2005; Clarke, 2008; Collier, 2012; Ferguson, 2010). Neoliberalism has become something of a "concrete abstraction" in academic discourse (Comaroff, 2011). According to Rachel Turner, what the concept "stands for and what it explains is both confused and confusing" (2008: 2). The danger, therefore, is that neoliberalism merely serves as a catchall term for all sorts of things. In his wide-ranging work on pragmatism philosopher Richard Rorty has written about the importance of actively seeking to overcome what he labels "entrenched vocabularies" (1989: 8). The implication of Rorty's statement is that extant conceptual vocabularies, having become ingrained and institutionalized, impede the development of new forms of thought. The case of neoliberalism is again instructive here: there has been a tendency for researchers to see the concept through an entrenched critical prism that somehow occludes intentionality, and consequently they lose sight of the stubborn wills operating in the background. As such, there is something to be said about paying attention to the politics of neoliberal processes (see both Howard and Shields in this volume), to their generative capacities, and to how they are linked with novel techniques of government. While neoliberalism may have been over-theorized of late, as a concept it nevertheless has a haunting presence in everyday life. This presence can be apprehended by moving away from stylized modes of critique to a type of re-description that is congruent with Rorty's approach. The ethnographic turn within studies of governmentalities can be seen in a sense as a vehicle for putting Rorty's ideas to work.

The ethnographic turn has thrown considerable light on changing practices of government in neoliberalizing contexts (see Li, 2007). Rather
than attempt to debunk the critique of neoliberalism as such, we wish
to highlight the practices embodied in "the public, shared vocabulary
we use in daily life" (Rorty, 1989: xv) in making sense of neoliberalism.

This is where the concept of governmentality comes into play, for
it has a unique relationship with neoliberalism. Indebted to the work
of Michel Foucault, a central analytical thrust of the governmentality
literature is the concept of government. One of Foucault's most reso-
nant ideas, government is more or less a descriptor of strategic intentionality (Foucault, 2000a). It is strategic insofar as it revolves around
attempts by authorities of various kinds to achieve objectives through
calculated means, albeit with a view towards "the right disposition
of things" (ibid.: 208). In Foucauldian terms, the right disposition of
things is shorthand for arranging things in a careful and exacting manner so as to lead to a specific end. The right disposition of things aligns
with a prevailing neoliberal logic that involves attempts to reorganize
society in response to historically specific economic, social, and political problems. Arranging society in this manner plays out through the
art of governing human conduct, which entails directing the "possible
fields of action of others" (Foucault, 1982: 221). Calculation is central
in this scheme of arrangement, and on the whole it is directed towards
specific finalities in the pursuit of what is deemed to be the common
good (Foucault, 2000a: 211).

Beyond its conceptual utility, the governmentality approach also
has methodological value. Foucault suggested that the analytics of
governmentality amounted to "a method of decipherment" (Foucault,
2008: 186). Nikolas Rose, a key interpreter of Foucault's œuvre, also
refers to the "ethos of analytics of governmentality," contending that
governmentality is grounded in empirical realities. In the preface
to the second edition of his groundbreaking *Governing the Soul*, Rose
interprets Foucault's methodological procedure as entailing concrete
investigations of the forms of rationality constituting present conditions (see Rose, 1999a: vii–xxvi), and we suggest that this approach also
should include explorations of what Foucault described as a "witches'
brew" of practices (2000b: 232). These practices, while underwritten by
largely instrumental conceptions of means and ends, transpire in specific places and at certain times. Studies of governmentality, according
to Rose, "are studies of a particular 'stratum' of knowing and *acting*"
(1999b: 19; emphasis added). However, this perspective, is lost in most

appropriations of governmentality and thus elides any serious consideration of what sort of action can be exercised by authorities of diverse kinds – on willing and unwilling subjects alike. ("actors")

It is this focus on governmentality as methodology that allows us to examine a particular moment in a two-decades-long treaty process in British Columbia, Canada, involving many First Nations and the governments of British Columbia and Canada. We focus on a related province-wide treaty referendum and associated public consultation in 2001–2 (Rossiter & Wood, 2005) and explore how dual logics of "certainty" and publicness were transformed into a medium for doing politics in a governmental register. We draw on hitherto neglected transcripts of province-wide consultation hearings held in 2001 by the provincial Select Standing Committee on Aboriginal Affairs in fifteen communities across British Columbia, which paved the way for the treaty referendum, which was held in early 2002. We use the treaty referendum process to analyse contemporary contexts of neoliberal governmentalities that emerge in relation to the unique social and political-economic relations of land. The treaty referendum also enables us to draw lines of connection between neoliberal governmentalities and what Foucault famously described as "the conduct of conduct" (1982: 220–1), highlighting the ties that bind the practice of politics and the practice of public mediation. As such, this chapter represents an ethnography not just of land, but of neoliberal governmentality. Thus, our main objective here is to draw attention to the use of the treaty referendum by the BC government to create synergies between the strategic objectives of the province and the agency of the electorate in relation to the prickly "land question" that has persistently dogged the province. To be sure, the land question has not arisen in a vacuum. Rather, it owes its provenance to the fact that Aboriginal rights and title in British Columbia pose a legal burden on Crown title, with implications for economic certainty over lands in the province.

By seeking to mobilize the agency of members of the public in achieving governmental ends of "certainty" and publicness, the treaty referendum process illuminates the work that "ordinary people" are meant to perform in neoliberal times as they are effectively "summoned as partners or participants in new assemblages of power" (Newman & Clarke, 2009: 46). Within a governmentality framework overlapping and contradictory governmental rationalities intersect with categories such as responsibilization (see also Larner in this volume), with its emphasis on skills of participation and responsible action (Shamir, 2008). Governing through responsibilization thus entails a neoliberal emphasis on individual responsibility, or the individualization of responsibility. In this move, the very nature of individuality is transformed along the lines of agency. In the contemporary world responsibilization plays out in efforts to enrol ordinary citizens as active agents in the task of fashioning governmental solutions to defined problems. The treaty referendum process can be seen as constitutive of a distinctive habitat of neoliberal responsibilization, one that is emblematic of a thoroughgoing reordering of the balance of responsibility between the state and citizens. Ronen Shamir's definition is instructive here; for him, responsibilization is "a call for action; an interpellation which constructs and assumes a moral agency and certain dispositions to social action that necessarily follow" (ibid.: 4). Seen in these terms, responsibilization functions as "an 'enabling praxis' and a technique of government" (ibid.).

Our analysis builds to some extent on arguments claiming that governmentality has a lot to offer for the analysis of society in a neoliberal age. But we are also concerned with developing a focus on the connections between contestation and the rise of neoliberal governmentalities. In analysing the treaty referendum process we adopt Rose's methodological axiom: "To analyze ... through the analytics of governmentality is not to start from the apparently obvious historical or sociological question: what happened and why? It is to start by asking what authorities of various sorts wanted to happen, in relation to problems defined how, in pursuit of what objectives, through what strategies and techniques" (1999a: 20). The essential thrust of Rose's injunction here is that the analytics of governmentality enable researchers to track aspirations that are folded into specific discourses, practices, and fields of strategic interventions that link together a range of motivations, aspirations, and desires. In effect, what authorities *want* to realize is shorthand for strategic intentionality, a concept we are particularly keen to employ in this chapter.

In what follows, the chapter detours through key literatures on governmentality and the larger corpus of Foucauldian writings on government and responsibilization to explore the extent to which aspects of Foucault's ideas can be put to work to analyse the politics involved in governing through land in the BC treaty process.

The Land Question in British Columbia: A Haunting Presence

Through the conceptual prism of governmentality, we approach land as a field of political manœuvrings and a tool of power, including the government of conduct, with a view towards "the right disposition of things" (Foucault, 2000a: 208). In this way we aim to analyse the politics of neoliberal governmentalities taking shape within the postcolonial contestation of British Columbia in the form of a treaty process (see Egan, 2012, 2013; Price, 2009; Woolford, 2005). As Li observes, "with land ... it is never over" (2014: 591). The chapter therefore relies on a framing of land as a lively ethnographic object – the locus of grounded and often conflicted human practices that unfold in time. Land is implicated in and productive of a dense network of relations, which are not merely about who owns what portion of earth but are more fundamentally linked to ideas of being in the world. As such, our views of land should not be limited to material realities alone, but placed within the framework of a host of powerful discursive and practical devices. By constituting the basis for everyday human interaction, land sets the framework for how space is to be occupied, appropriated, and shared, generating a slew of "consequential geographies" (Blomley, 2005: 127). Put simply, land is essential for human activity, yet it is also intrinsically rivalrous. Not only does it mean different things to different people, but it also tends to be non-substitutable. Seen this way, land represents a complex of issues, problems, and objects that generate contention, serving as a focus for the formation of competing public imaginaries and helping to sustain all sorts of political, legal, and cultural claims. In short, land is the basis for struggles – active or passive – concerning the right to define, appropriate, use, or benefit from it.

Conflict over land as a particular, demarcated, territorial formation (and all that this entails) was and still is central to colonialism. In postcolonial contexts land has become the nodal point of debates of critical political and practical importance regarding the structuring of social relations between settlers and Indigenous populations. The constraints and possibilities of land control have played out in multiple spaces of settlement around the world – from Canada to South Africa, the United States to New Zealand. Settlement of these new frontiers by Europeans depends on the opening of territory that can be physically and legally possessed. This aspect is intensified in a settler society such as British Columbia, where the provincial economy depends on access to resources (lumber, minerals, etc.) dispersed across a far-flung terrain.

As many other scholars have noted, British Columbia was resettled on a premise not dissimilar to the doctrine of *terra nullius* that underwrote the resettlement of Australia. This legal doctrine held that land previously settled by Indigenous groups was unencumbered and bereft of title and therefore free for European resettlement (see Blomley, 2003; Day & Sadik, 2002). In many ways, terra nullius actively shaped the colonial endeavour of the late nineteenth century and defined the power of the colonial state. But what is striking is that the case of British Columbia is atypical in Canada because, unlike other parts of the country, the ingrained belief in terra nullius had far-reaching consequences for colonial settlement. Between 1850 and 1854 James Douglas, the first colonial governor of British Columbia, had negotiated fourteen treaties with Indigenous groups on Vancouver Island on behalf of the British Crown. But the so-called Douglas Treaties were far from extensive, covering only small areas (Harris, 2002: 19), and subsequent provincial governments abandoned treaty making, on the principle that Indigenous communities had no title with which to negotiate. This was the position held by British Columbia when it entered the Canadian confederation in 1871. Under the Terms of Union, British Columbia was given complete jurisdiction over lands and resources, and lands designated as public passed into Crown ownership.

Over a century and half later the "land question" continues to maintain a spectral presence, haunting the province in all sorts of ways. It took on a judicial dimension in 1973 when the Nisga'a took their land rights case to the Supreme Court of Canada. The court's judgment in the Calder case led to a shift in the federal approach to the "Indian question," and Prime Minister Pierre Trudeau's government took a decidedly different stance regarding Aboriginal rights. Trudeau approved the Constitution Act of 1982, which affirmed the pre-existence of Aboriginal title. Although the federal government had acknowledged continuing Aboriginal title in the 1973 Statement of Claims of Indian and Inuit People, the government of British Columbia continued to deny it.

The 1980s heralded a shift in the province's strategy of refusing to negotiate modern-day treaties or give recognition to Aboriginal title. A series of militant acts on the part of First Nations in the form of blockades occurred in places such as Meares Island, situated in Clayoquot Sound on the west coast of Vancouver Island, and Haida Gwaii, an archipelago off the north coast of the province. The blockades went hand in hand with legal action to prevent resource extraction in Aboriginal territories (see Blomley, 1996; Ramos, 2006; Rossiter, 2014). Against this backdrop of events, the provincial Social Credit government established a Ministry of Native Affairs in 1988 to cater to Aboriginal concerns, while adhering to the provincial policy of not negotiating treaties. However, there was a growing conviction in governmental and public quarters that non-negotiation was unsustainable (Tennant, 1996: 55). In July 1989 the Native Affairs Advisory Council was formed by then Premier Bill Vander Zalm to explore the unresolved land question. A much publicized tour of the province in late 1989 and early 1990 by Vander Zalm and members of the council led to meetings with leaders of the First Nations councils, in which Aboriginal demands were discussed (ibid.: 56). Following recommendations put forward by the council, the provincial government announced its intention to enter into negotiations with First Nations. To this end, the BC government and the First Nations Congress established a task force aimed at kick-starting the process of negotiations in the province.

In December 1990 the British Columbia Claims Task Force was created, with the overarching goal of putting forward recommendations for negotiated solutions for the Aboriginal land question. On 28 June 1991 the report of the task force was released (BC Claims Task Force, 1991). It comprised nineteen specific recommendations for resolving land disputes in British Columbia regarding the use, administration, and control of land and resources. The task force recommended the creation of a six-stage treaty process in which Aboriginal groups could participate of their own accord. In 1992 representatives of the First Nations Summit and the federal and provincial governments made a formal commitment to a treaty process by signing the BC Treaty Commission Agreement at a public ceremony at the Squamish Nation. With the signing of the agreement the BC Treaty Commission (BCTC) was established to coordinate the negotiation of treaties.

Two significant events intensified the politics of land in British Columbia in the late 1990s. On 11 December 1997 the ruling of the Supreme Court of Canada in the landmark case *Delgamuukw v. British Columbia* further defined what was at stake in the treaty negotiations process. The Supreme Court ruled that Aboriginal title was a right to land, and that at no time was this title ever extinguished in British Columbia. Shortly thereafter British Columbians awoke to headlines declaring that "BC's native Indians are laying claim to every tree, every rock, every fish and every animal in the province" (Ouston, 1998: A1). "Everything's up for grabs," declared a conservative publication, expressing worry at a "looming free-for-all" (Thompson, 1999). The Nisga'a Final Agreement, negotiated outside the BC Treaty Commission process, was signed in August 1998 between the governments of Canada and British Columbia and the Nisga'a Nation of northwestern British Columbia. Heralded widely as the province's first modern treaty, upon ratification in

2000 this agreement granted to the Nisga'a people, among other things, 1,930 square kilometres of land in the lower Nass Valley. The Nisga'a agreement was attacked by conservative actors, who saw it as creating an intolerable space of difference (of race, sovereignty, property, rights) at the centre of a unitary polity.

Pandering to popular anxieties at the fracturing and fast-moving "edge politics" (Howitt, 2001) of the day, the provincial opposition leader, Gordon Campbell, declared the Nisga'a Final Agreement "unconstitutional," launching a lawsuit in October 1998 that challenged its validity. Two years later in *Campbell et al. v. AG BC/AG Canada & Nisga'a Nation et al.*, the BC Supreme Court upheld the constitutionality of the agreement. When the provincial general elections of 2001 resulted in a Liberal government and his becoming premier, Campbell announced his intention of holding a province-wide referendum on treaties in British Columbia, in keeping with the right-of-centre logic of direct democracy via plebiscites and recall measures (see Miller, 2009; Penikett, 2006). During that year the government initiated the promised referendum, which was eventually held between 2 April and 15 May 2002.

Enrolling the Public, Mobilizing Certainty

A good starting point for an analysis of the treaty referendum is a 27 April 2002 article in the *National Post* in which Premier Campbell articulated his government's justification for the process:

> We believe it's time the *public* was included in this process. The referendum will give British Columbians a direct say on the principles that we believe should guide the province's approach to treaty negotiations. It will provide *certainty* for the province's negotiating position. It will reinvigorate the treaty process. And it will build a foundation for a new relationship with First Nations that ensures Aboriginal British Columbians share fully in a prosperous future.
>
> The referendum is an opportunity for all of us to understand our constitutional obligations to Aboriginal people, and discuss how we can move forward with settlements in a way that has *public* trust and confidence – to help fast-track treaty negotiations, and forge a new relationship with First Nations. Fundamentally, the referendum is about seeking and honouring the will of the people. It is British Columbia's first provincial referendum in over a decade, and one of the most comprehensive in Canada, in terms

of presenting a detailed series of public policy questions to be answered individually. (2002; emphasis added)

Campbell's own framing of the scope of the referendum is fairly innoc-uous. Rather than assuming that members of the public do, in fact, have a fully formed idea of what the treaty process entails, the objective here is to educate the public by raising awareness of the issues and obliga-tions at stake. In this regard, the notion of the public serves as a tool of government. It therefore takes on particular importance in the conduct of conduct. In *Security, Population, Territory* Foucault highlighted the instru-mental value of the public, defined as "the population seen under the aspect of its opinions, ways of doing things, forms of behaviour, customs, fears, prejudices and requirements; it is what one gets a hold on through education, campaigns and convictions" (2007: 75).

More interestingly, when read along and against the grain, Camp-bell's words are underwritten by an instrumental logic of individual responsibilization that is framed in part around the imagery of pub-lic participation. In this moment of responsibilization members of the "public" are positioned as democratic agents engaged in deliberation and exploration while simultaneously framed as productive of finality. Campbell's solicitation of ordinary people as agents of certainty is also intimately entangled with the reasoning that the greater good of the province is at stake. This is hardly surprising, given that Campbell had advanced the goal of economic certainty during his electoral campaign. Reference to the "will of the people" frames public action as in itself constitutive of the possibilities of active individuals acting together to achieve a common purpose, albeit one in which his government was invested in.

Similarly, in a news release from the Treaty Negotiations Office, BC Attorney General Geoff Plant also articulated his thoughts on the basis for the referendum: "We want to negotiate lasting agreements that are based on a clear sense of purpose and direction from the electorate of British Columbia. We are encouraging voters to participate in the ref-erendum so that we establish a publicly supported mandate that will help us move forward and achieve tangible results in the treaty pro-cess" (BC Treaty Office, 2002).

In the above extract Plant invokes and inserts into the discourse surrounding the referendum what is perceived to be a commonsensi-cal and unproblematic universalist image of "the people" as a distinct type of public. Janet Newman and John Clarke have argued that the construction of any object as a matter for the public "involves political struggles to make them so" (2009: 2). In short, there is some level of effort involved and this effort is shaped by political intents. It is the very act of addressing the people that creates them as a social category – a public. In other words, the public exists only because of their enactment as such. The ability to pull off this summoning act lies in the extent to which "an addressable object" can be conjured into being (Warner, 2002: 97).

Within the reasoning framework of the architects of the treaty referendum, the greater good of the province lay in moving the contentious treaty process forward with a "clear sense of purpose and direction from the electorate of British Columbia" (BC Treaty Office, 2002). This vision worked itself out through intertwined processes of enrolment, facilitation, and responsibilization by the Select Standing Committee on Aboriginal Affairs. This committee was tasked with overseeing publicly held consultations on the form and scope of questions for the referendum on "principles" that could potentially inform future treaty negotiations. Comprising only Liberal party members of the Legislative Assembly (MLAs), the ten-member committee was chaired by Chilliwack-Sumas MLA John Les. The committee held hearings around the province between October and November 2001 and also solicited written submissions from willing members of the public as well as interest groups, including local government representatives and organizations such as the Anglican Church, the Council of Canadians, and the Council of Forest Industries. Textual devices were deployed as media of enrolment. The committee placed advertisements in various local and regional newspapers, and people were asked to contact the Office of the Clerk of Committees to include their names in hearing agendas. Through a publicly available website interested individuals were given access to materials relating to the committee's mandate. In a practical sense, the public were enrolled twice: first through the hearings, and second through the referendum itself.

In a 21 September 2001 news release made available a few weeks before the start of the hearings Les reiterated the BC government's vision for the committee's activity: "The goal of the committee is to listen to all British Columbians. We want to engage the public in a constructive and useful discussion that will reinvigorate the treaty negotiation process in this province. Negotiating effective treaties with First Nations is a commitment of the government, and provincial principles guiding negotiations will be strengthened through public input" (BC Legislative Assembly, 2001a).

In a sense, the committee was tasked with constructing a set of concerns around the treaty negotiation process and devising appropriate solutions. For Les, public input was vital to a credible solutions framework. At the same time he stressed that, in spite of criticisms that the looming referendum was nothing more than a tool of political and economic self-interest, the public consultation process was a thoroughly democratic one: "We think the referendum will be a part of the solution as opposed to a problem" (BC Legislative Assembly, 2001b: 105).

Some insight into the assumed benefits of public participation can also be gleaned from the following excerpt from a 19 September 2001 presentation in Victoria by Deputy Minister Phil Steenkamp of the Treaty Negotiations Office:

> One of the objectives of the referendum, as I understand it, is to provide for public input into the principles which will inform the provincial government's mandate at the table and, beyond that, to really give British Columbians a sense that they have had a say, that they have had a voice, and to legitimize provincial participation in treaty-making. If government is successful in that mission, we should, at the end of the referendum process, have a public which has had its say on the questions put in front of it. (BC Legislative Assembly, 2001c: 35)

The referendum hearings, therefore, enrolled the public in what appeared to be a wide-ranging program of democratic deliberation on the treaty process, potentially opening up the black box of settler colonialism and the multiple entanglements of land. Yet, at the same time, the hearings and subsequent referendum were expected to produce something called "certainty." As MLA Dennis MacKay, who at the time represented the electoral district of Bulkley Valley-Stikine, put it during a hearing in Prince Rupert on 3 October 2001: "Economic development in British Columbia today, I would suggest, is somewhat stifled because of the uncertainty over land claims issues. If we can move this process on and come to an end date down the road – and I hope it's sooner rather than later – the economic benefit is going to benefit you people, the native people, and the non-native people. It's going to benefit everybody" (BC Legislative Assembly, 2001b: 104).

A straightforward critique of this discursive manœuvre is that it is parasitic on existing predispositions of the BC public. But this interpretation tells only half the story. Analytics of governmentality make it possible to argue that certainty is in fact being redefined in talk as a realm through which the public can demonstrate their personal commitments to the wider instrumental logic of treaty making. In these examples, the characterization of the "public" as a set of rational and responsible actors is a classic neoliberal practical move that relies on ideas of choice, freedom and agency (see Rose, 1999b). "Certainty" is a strange, and in many ways, uncertain beast, that does complicated work (Blackburn, 2005; Woolford, 2005). Notionally, certainty refers to a "legal technique that is intended to define with a high degree of specificity all of the rights and obligations that flow from a treaty and ensure that there remain no undefined rights outside of a treaty" (Stevenson, 2001: 114).

In that sense, the mobilization of certainty in official discourse is strongly aligned with a discourse of finality and temporal closure. Yet "certainty" also appears to be a placeholder for the wide-ranging and deeply ingrained anxieties that the land question and Aboriginal rights evoke in settler society. As a leitmotif, it continues to be almost obsessively invoked, as evidenced by a federal factsheet on the BC treaty negotiations:

> *Uncertainty* about the existence and location of Aboriginal rights creates *uncertainty* with respect to ownership, use and management of land and resources. That *uncertainty* has led to disruptions and delays to economic activity in BC. It has also discouraged investment.
>
> The consequences of not concluding treaties will be lost economic activity as well as escalating court costs and continued *uncertainty*. Key benefits of negotiated settlements are economic and legal *certainty* as well as harmonized arrangements between the different levels of government. (Aboriginal Affairs and Northern Development Canada, 2010; emphasis added)

It is hardly surprising, then, that certainty has had a defining presence in the political terrain of the treaty process. At its most basic, certainty was presented by the BC government in a commonsensical register, framed as a desirable and incontrovertibly beneficial concept generating, among other things, economic security and stability. This particular framing of certainty rests on the traditional view held by successive BC governments that unresolved Aboriginal rights and title impeded business interests in the province's land-based resource economy. On this score certainty was held up as a means through which the interests of First Nations, the government, and the private sector could be properly aligned. This view is captured in the following statement by Attorney General Geoff Plant, who was the minister responsible for treaty negotiations: "Negotiated settlements with First Nations ... represent an opportunity to achieve certainty over lands and resources in the region – an opportunity that our government has identified as a priority so that we can attract investment and create more economic stability in BC" (cited in Ponting, 2006: 135).

Plant's point about certainty here is succinct and to the point. But looked at more closely, it seems to conform to a broader rationality that holds that certainty is an achievement – an exercise in arranging things for instrumental ends. The same kind of rationality seems to underlie a November 2002 cabinet address in which Plant articulated the BC government's strategy for achieving this objective: "Certainty lies in the strength and stability of the entire treaty, so we're not looking for just one single method to achieve certainty. We think we can achieve certainty in a *variety of arrangements* that correspond to particular issues or circumstances" (BC Treaty Commission, 2003: 3; emphasis added). This is a long-standing concern. The province's position paper, defining its negotiation mandate, specified: "Treaty negotiations will exchange ... relatively *undefined* Aboriginal rights with *clearly defined* rights to land and resources in a manner that *fits* with contemporary realities of economics, law and property rights in British Columbia" (British Columbia, 1995; emphasis added).

In the above excerpts certainty emerges as a bounded and "intelligible field with specifiable limits and particular characteristics" (Rose, 1999b: 33). This field is intimately linked to solutions and explicit programs of intervention that fall within the repertoire of governmentality. Certainty arranges a number of things. Most immediately, it seeks to make legible and hook up a topography of economic units (Blomley, 2015), through "a variety of arrangements." Connection goes hand in hand with severance, however, in the arrangement of things. Certainty, working through "contemporary realities," organizes time, drawing a veil over the past and bracketing the future (Rossiter & Wood 2005). Treaties are not intended, argues the Crown, to achieve redress for past wrongs. They draw a line under the past and allow us to move forward together. However, in so doing they configure reconciliation as a specific and final moment, that ends with signatures, rather than thinking of the treaty, as do many First Nations, as "an ongoing activity, a continuous process of cross-cultural dialogue over time" (Tully, 2001: 13).

The Select Standing Committee on Aboriginal Affairs delivered its final report to the provincial legislature on 30 November 2001. The report sought to weave together 482 written submissions and 15 public hearings. Following the release of the report, Attorney General Geoff Plant tabled a motion to hold the referendum in the spring of 2002 under the provincial Referendum Act. In the end eight referendum questions were posed in a mail-in ballot to the BC electorate (Elections BC, 2002: 9). Each question hinged on individual "principles" that could inform future treaty negotiations, for which an unequivocal "yes" or "no" answer was required. The referendum ballots were sent by mail to registered voters by Elections BC, the office of the BC Legislature responsible for conducting elections in the province. Although participation rates were low,[1] Premier Campbell claimed a victory in a post-referendum press conference held on 2 July: "Today's referendum results mark a crucial milestone in the long march towards treaty making in British Columbia. After many years of being shut out of the treaty process, *the people* have finally had their say – and their message to First Nations and to all Canadians is unmistakable. British Columbians stand firm in their resolve to negotiate workable, affordable treaties that will provide *certainty*, *finality* and equality. They have given their provincial government a *clear mandate* and a solid set of principles to get on with the task" (British Columbia, 2002; emphasis added).

At the most general level this framing of the results of the referendum as empowering not only to society as a whole but also to the government represents an effort to construct the referendum as a medium of bottom-up decision making (although, as far as we can tell, the practical effects on the negotiating position of the province were minimal). But more important, the excerpt generates a dynamic in which Campbell uses practical modes of reasoning to infer a relationship between governmental objectives and outcomes. In this sense, the attribution of choice to the people does the work of rendering the referendum and the continued treaty process valid, without any further necessary justification.

Recomposing Neoliberal Governmentalities?

The strategy of struggle also constitutes a frontier for the relationship of power, the line at which, instead of manipulating and inducing actions in a calculated manner, one must be content with reacting to them after the event ... In effect, between a relationship of power and a strategy of struggle there is a reciprocal appeal, a perpetual linking and a perpetual reversal.

(Foucault, 1982: 225–6)

So far we have harnessed insights from the literature on neoliberal governmentality to develop our analysis on the politics of certainty, which is an offshoot of the lingering land question in British Columbia. More specifically, we employed the concept of governmentality to tease out the strategic intentionality behind the 2002 treaty referendum. In effect, we have located the treaty referendum within a wider context characterized by a new modality of politics distinguished by powerful discursive and practical rationalities that seek "a whole series of specific finalities" (Foucault, 2000a: 211).

At the same time we would like to suggest that studies of neoliberal governmentalities do not pay sufficient attention to interactive dynamics that may emphasize differential logics and the efforts of a broad range of actors pursuing plural ends. In other words, we want to urge a shift away from the reasoning that the intentions of the state can only come off as conceived. Although the strategic impulse of governmentality may well hold sway, it does not totally displace elements of contestation (whether regarded positively or negatively). In putting governmentality ideas to work in empirical analysis, therefore, it is important to look beyond what authorities wanted to achieve. Although our primary goal in this chapter has been to examine the practical workings of the political rationalities of the 2002 treaty referendum, we also wish to displace the totalizing sense of agency ascribed to the state, specifically what it can achieve by mobilizing the public. During the treaty referendum hearings members of the public contested the logic underpinning the process, casting serious doubt on its overarching aims and criticizing the solutions proposed. In so doing the very members of the public summoned to validate the policies of the government have in turn provided sound reasons to be sceptical of some claims made in their name. They articulate what Andrew Sayer refers to as "lay normativity" or the "range of normative rationales, which matter greatly to actors, as they are implicated in their commitments, identities and ways of life" (2005: 6). According to Sayer, this dimension of lay normativity allows us to avoid seeing subjects as "bloodless figures who seemingly drift through life, behaving in ways which bear the marks of their social position and relations of wider discourses, disciplining themselves only because it is required of them, but as if nothing mattered to them" (ibid.: 51). Lay normativities allow people to differently interpret and apply norms in ways that lend themselves to interpretation in terms of refusal, contestation, bargaining, and so forth. They also provide pathways for individuals to develop alternative and persuasive arguments to support their opposition to dominant normative ideas. It is precisely lay normativities that redefine neoliberal governmentality as an open-ended register that enables the mobilization of differential norms, dispositions, and structures of feeling.

Although nearly all the hearings afforded members of the public an opportunity to make suggestions regarding referendum questions, either verbally or in written submissions, many participants expressed reservations about the need for a referendum process. In a presentation in Chilliwack on 17 October Victoria Jordan, a member of the Haida Nation, stated: "I have my doubts about the effectiveness of a treaty referendum in general. I fear that the general electorate of BC might view it as a cure-all to make everybody happy on both sides, when the reality for me is that the treaty process must continue anyway, regardless of the outcome of any referendum. Maybe we should be asking ourselves what this is really going to accomplish" (BC Legislative Assembly, 2001d: 300). This is a rather different interpretation of the governmental narrative of certainty and finality. In fact, in this case we see evidence that the Campbell government is hardly speaking for the public. One could argue that enabling ordinary voices redirects attention to the inadequacies of the treaty referendum process.

Some participants at the hearings sought to underscore the importance of allowing Indigenous voices to shape the conversation. Ernie Freeman, a social work practitioner, addressed this issue during the hearing in Fort St John on 5 October: "One of my big concerns was how to ensure that we incorporate the world view of Aboriginal nations within the treaty process ... I think this is where you need a strong Aboriginal voice in order to bring the other world view into this. I think you're not going to find either fully, like a dichotomy of totally transferring the economic model, a Eurocentric model, into a First Nations culture" (BC Legislative Assembly, 2001e: 199, 200). Here the "land question" receives a different inflection. As Li (2014: 589) notes, land "has an especially rich and diverse array of 'affordances'–uses and values it affords to us, including the capacity to sustain human life." The "land question" entails not only a question concerning ownership of the land, but also, it is suggested, a contention concerning the very status of "land" itself.

In the same vein, in a presentation at the Victoria hearing on 2 November Murray Browne, a lawyer representing Aboriginal groups, pointed to the one-sidedness of the dominant certainty argument: "I ... want to note that it's probably not fair for the province to demand certainty from First Nations when the province isn't offering certainty to First Nations. For example, the province has refused to incorporate environmental protection standards in treaties, and the province has refused to make commitments not to double-tax treaty settlement land. If we want to talk about fairness, if the provincial government wants certainty from First Nations, it should be offering certainty to First Nations" (BC Legislative Assembly, 2001f: 599).

Certainty, as an arrangement of things, is complicated in this account. In asking the province to offer certainty to First Nations, rather than vice versa, the submission invites a recalibration of the prevailing view that Aboriginal title is the burr under the saddle, the impediment to the smooth functioning of a given order. We are reminded of the comments of Frank Calder, in whose name the landmark 1973 case noted above was brought: "there is no such thing as an Indian land claim. He's got nothing to claim, we own the territory. It should have been called 'white land claims'" (cited in Foster, Rowen, & Webber, 2007: 53). Here, we see the land question in British Columbia being reframed as one that exceeds narrow frameworks of certainty. At work are alternative imaginaries, vocabularies, and processes of indigeneity that can be seen through the lens of a broader philosophical debate on Indigenous ontologies (see Blaser, 2014; De la Cadena, 2010; Law, 2004). These ontological formations are generally seen to pose a serious challenge to deeply enshrined western ways of categorizing and relating to land.

Conclusion

The proliferation of neoliberalism in academic discourse and, to some extent, public debate has not, of course, gone uncontested. The limitations of the prevalent conceptualization of neoliberalism have been exposed by way of all sorts of criticisms and interventions of late, and this trend has clearly been productive, not least because it has led to a concerted move towards unthinking "neoliberalism." Against that backdrop, this chapter has adopted a different analytical approach. We have sought not to reify neoliberalism by ascribing too much value to its nature, features, or characteristics, a focus we believe does nothing to disturb assumptions about neoliberal orthodoxies or to highlight their variegated forms. Yet, at the same time, we are of the view that a heavy focus on critiques of neoliberalism as of late has perhaps systematically obscured what is most distinctive about the ways in which the subjective meanings associated with the concept have been built up historically, shored up, and ascribed powerful rationalities. As we show in this chapter, the rationalities of government draw attention to a highly creative, yet shifting, set of discursive assemblages and practical registers through which certain normative neoliberal ideals play out. These registers are noteworthy because they appear to be increasingly reconfigured as a dimension of civic engagement and fuelled by compelling conceptions of the social good and public interest.

In this chapter we have employed the case study of the treaty referendum event of 2002 in British Columbia to conceptualize land as the locus of grounded human practices, couched within distinctive, yet fluid, modes of governmentality and framed by a politics of public action centred around certainty. In so doing we have attempted to give the term "ethnography" a particular orientation through a focus on concrete manifestations of government that bind together land and human subjects. By analysing land as a lively ethnographic object we simultaneously foreground the political agency of practices, including those that arise from some combination of neoliberal modalities of governmentality, as they unfold in time. In *Security, Population, Territory*, Foucault drew further attention to the extent to which neoliberal governmentality works through the people as a distinctive type of public. At the same time he highlights the extent to which populations link up with spatial rationalities, generating particular spatial configurations informed by attempts to appropriate spaces for different ends. Invoking the notion of "milieu," Foucault suggests that as a medium of action space involves a mix of natural and artificial "givens." It is precisely a "field of intervention" that not only affects individuals as legal subjects capable of voluntary actions but also as performative subjects (2007: 21).

At a particular level of abstraction, the milieu is a conceptual space of governmentality wherein the potential of conduct is harnessed. But more practically, the notion of milieu shines a light on the entangled space of social and biosocial relations, including land and its affordances in everyday life. Indeed, as Foucault has shown, government is a project that revolves around the assemblage of "men and things," including "relationships, bonds, and complex involvements with things like wealth, resources, means of subsistence, and, of course, the territory with its borders, qualities ... and so on" (ibid.: 96). The referendum process as we have described it so far represents a complex field of neoliberal governmentality in that it sought to connect the people with the land. By appealing to the voluntary actions of the citizenry, the architects of the referendum process sought to mobilize the agency of a narrowly defined public for the specific end of achieving certainty and finality with respect to the land question in British Columbia. The public is analysed in this chapter along broadly consequentialist lines of mediation and responsibilization, which in this case represents one vector of a distinctive strategy of government.

NOTE

1 By the deadline of 15 May Elections BC had received 35.84 per cent of the ballots (763,480). According to the results indicated, voters gave mostly "yes" responses (between 84.5 and 94.5 per cent) to the eight questions (Elections BC, 2002: 6).

REFERENCES

Aboriginal Affairs and Northern Development Canada (2010). Fact sheet – Treaty negotiations. Available at https://www.aadnc-aandc.gc.ca/eng/1100100032288/1100100032289. Accessed 20 March 2015.

Barnett, C. (2005). The consolations of "neoliberalism." *Geoforum*, 36(1), 7–12. http://dx.doi.org/10.1016/j.geoforum.2004.08.006

BC Claims Task Force (1991). Report of the British Columbia Claims Task Force. Available at http://www.fns.bc.ca/pdf/BC_Claims_Task_Force_Report_1991.pdf. Accessed 6 September 2011.

BC Legislative Assembly (2001a). News release: Public hearings to begin on treaty referendum questions. http://www.leg.bc.ca/cmt/37thparl/session-2/aaf/media/releaseSept21.htm. Accessed 7 September 2011.

BC Legislative Assembly (2001b). Minutes, Select Standing Committee on Aboriginal Affairs, Wednesday, October 3, 2001, Prince Rupert, BC. Available at https://www.leg.bc.ca/cmt/37thparl/session-2/aaf/hansard/a20011003p-final.htm. Accessed 7 December 2011.

BC Legislative Assembly (2001c). Minutes, Select Standing Committee on Aboriginal Affairs, Wednesday, September 19, 2001, Victoria, BC. Available

at https://www.leg.bc.ca/cmt/37thparl/session-2/aaf/hansard/
a20010919a-final.htm. Accessed 7 December 2011.

BC Legislative Assembly (2001d). Minutes, Select Standing Committee
on Aboriginal Affairs, Wednesday, October 17, 2001, Prince Rupert, BC.
Available at https://www.leg.bc.ca/cmt/37thparl/session-2/aaf/hansard/
a20011017a.htm. Accessed 7 December 2011.

BC Legislative Assembly (2001e). Minutes, Select Standing Committee on
Aboriginal Affairs, Friday, October 5, 2001, Fort St. John, BC. Available at
https://www.leg.bc.ca/cmt/37thparl/session-2/aaf/hansard/a20011017a.
htm. Accessed 7 December 2011.

BC Legislative Assembly (2001f). Minutes, Select Standing Committee on
Aboriginal Affairs, Friday, November 2, 2001, Victoria, BC. Available at
https://www.leg.bc.ca/cmt/37thparl/session-2/aaf/hansard/a20011017a.
htm. Accessed 7 December 2011.

BC Treaty Commission (2003). Treaty Commission update: January 2003.
Available at http://www.bctreaty.net/files/pdf_documents/january03update.
pdf. Accessed 16 May 2012.

BC Treaty Office (2002). April 2, 2002 news release: Treaty referendum begins
today. http://www.gov.bc.ca/tno/news/2002/referendum_begins_today.
htm. Accessed 3 October 2011.

Blackburn, C. (2005). Searching for guarantees in the midst of uncertainty:
Negotiating Aboriginal rights and title in British Columbia.
American Anthropologist, 107(4), 586–96. http://dx.doi.org/10.1525/
aa.2005.107.4.586

Blaser, M. (2014). Ontology and indigeneity: On the political ontology of
heterogeneous assemblages. *Cultural Geographies, 21*(1), 49–58. http://
dx.doi.org/10.1177/1474474012462534

Blomley, N. (1996). "Shut the province down": First Nations blockades in
British Columbia, 1984–1995. *BC Studies, 3*(1), 5–35.

Blomley, N. (2003). Law, property, and the geography of violence: The frontier,
the survey, and the grid. *Annals of the Association of American Geographers,
93*(1), 121–41. http://dx.doi.org/10.1111/1467-8306.93109

Blomley, N. (2005). Remember property? *Progress in Human Geography, 29*(1),
125–7.

Blomley, N. (2015). The ties that blind: Making fee simple in the British
Columbia treaty process. *Transactions of the Institute of British Geographers,
40*(2), 168–79. http://dx.doi.org/10.1111/tran.12058

British Columbia (1995). British Columbia's approach to treaty settlement:
Lands and resources. Available at http://www.llbc.leg.bc.ca/public/
pubdocs/bcdocs/272760/context.htm. Accessed 11 January 2015.

British Columbia (2002). Information bulletin. Premier Campbell: Post-referendum press conference. Available at http://www2.news.gov.bc.ca/archive/2001-2005/2002OTP0018-000502.htm. Accessed 3 April 2012.

Campbell, G. (2002). Referendum will spur treaty process. Available at http://www.vcn.bc.ca/~dastow/npo20427.txt. Accessed 14 June 2012.

Clarke, J. (2008). Living with/in and without neoliberalism. *Focaal: European Journal of Anthropology, 2008*(51), 135–47. http://dx.doi.org/10.3167/fcl.2008.510110

Collier, S.J. (2012). Neoliberalism as big leviathan, or ...? A response to Wacquant and Hilgers. *Social Anthropology, 20*(2), 186–95. http://dx.doi.org/10.1111/j.1469-8676.2012.00195.x

Comaroff, J. (2011). The end of neoliberalism? *Annals of the American Academy of Political and Social Science, 637*(1), 141–7. http://dx.doi.org/10.1177/0002716211406846

Day, R.J.F., & Sadik, T. (2002). The BC land question, liberal multiculturalism, and the spectre of Aboriginal nationhood. *BC Studies, 134*(1), 5–34.

De la Cadena, M. (2010). Indigenous cosmopolitics in the Andes: Conceptual reflections beyond "politics." *Cultural Anthropology, 25*(2), 334–70. http://dx.doi.org/10.1111/j.1548-1360.2010.01061.x

Egan, B. (2012). Sharing the colonial burden: Treaty-making and reconciliation in Hul'qumi'num territory. *Canadian Geographer/ Géographe canadien, 56*(4), 398–418. http://dx.doi.org/10.1111/j.1541-0064.2012.00414.x

Egan, B. (2013). Towards shared ownership: Property, geography and treaty making in British Columbia. *Geografiska Annaler. Series B, Human Geography, 95*(1), 33–50. http://dx.doi.org/10.1111/geob.12008

Elections, B.C. (2002). *Report of the chief elections officer on the treaty negotiations referendum.* Victoria: Elections BC.

Ferguson, J. (2010). The uses of neoliberalism. *Antipode, 41*(s1), 166–84. http://dx.doi.org/10.1111/j.1467-8330.2009.00721.x

Foster, H., Rowen, H., & Webber, J. (Eds). (2007). *Let right be done: Aboriginal title, the Calder case and the future of indigenous rights.* Vancouver: UBC Press.

Foucault, M. (1982). Afterword: The subject and power. In H.L. Dreyfus & P. Rabinow (Eds), *Michel Foucault: Beyond structuralism and hermeneutics* (208–26). Chicago: University of Chicago Press.

Foucault, M. (2000a). Governmentality. In J.D. Faubion (Ed.), *Power: Essential works of Foucault, 1954–1984.* Vol. 3 (201–22). New York: New Press.

Foucault, M. (2000b). Questions of method. In J.D. Faubion (Ed.), *Power: Essential works of Foucault, 1954–1984.* Vol. 3 (223–39). New York: New Press.

Foucault, M. (2007). *Security, population, territory: Lectures at the Collège de France, 1977–1978.* Basingstoke, UK: Palgrave Macmillan. http://dx.doi.org/10.1057/9780230245075

Foucault, M. (2008). *The birth of biopolitics: Lectures at the Collège de France, 1978–1979.* Basingstoke, UK: Palgrave Macmillan. http://dx.doi. org/10.1057/9780230594180

Harris, C. (2002). *Making native space: Colonialism, resistance, and reserves in British Columbia.* Vancouver: UBC Press.

Howitt, R. (2001). Frontiers, borders, edges: Liminal challenges to the hegemony of exclusion. *Australian Geographical Studies, 39*(2), 233–45. http://dx.doi.org/10.1111/1467-8470.00142

Law, J. (2004). *After method: Mess in social science research.* London: Routledge.

Li, T.M. (2007). *The will to improve: Governmentality, development and the practice of politics.* Durham, NC: Duke University Press. http://dx.doi. org/10.1215/9780822389781

Li, T.M. (2014). What is land? Assembling a resource for global investment. *Transactions of the Institute of British Geographers, 39*(4), 589–602. http:// dx.doi.org/10.1111/tran.12065

Miller, J. (2009). *Compact, contract, covenant: Aboriginal treaty-making in Canada.* Toronto: University of Toronto Press.

Newman, J., & Clarke, J. (2009). *Publics, politics, and power: Remaking the public in public services.* London: Sage.

Ouston, R. (1998, 2 February). B.C. Indian chiefs lay claim to entire province, resources. *Vancouver Sun,* A1–A2.

Penikett, A. (2006). *Reconciliation: First Nations treaty making in British Columbia.* Vancouver: Douglas & McIntyre.

Ponting, J.R. (2006). *The Nisga'a treaty: Polling dynamics and political communication in comparative context.* Toronto: University of Toronto Press.

Price, R.T. (2009). The British Columbia treaty process: An evolving institution. *Native Studies Review, 18*(1), 139–67.

Ramos, H. (2006). What causes Canadian aboriginal protest? Examining resources, opportunities, and identity, 1951–2000. *Canadian Journal of Sociology, 31*(2), 211–34. http://dx.doi.org/10.2307/20058697

Rorty, R. (1989). *Contingency, irony, and solidarity.* Cambridge: Cambridge University Press. http://dx.doi.org/10.1017/CBO9780511804397

Rose, N. (1999a). *Governing the Soul.*2nd ed.. London: Free Association.

Rose, N. (1999b). *Powers of freedom: Reframing political thought.* Cambridge: Cambridge University Press. http://dx.doi.org/10.1017/ CBO9780511488856

Rossiter, D.A. (2014). The nature of a blockade: Environmental politics and the Haida action on Lyell Island, British Columbia. In Y.D. Belanger & P.W. Lakenbauer (Eds), *Blockades or breakthroughs? Aboriginal peoples confront the Canadian state* (70–89). Montreal: McGill-Queen's University Press.

Rossiter, D.A., & Wood, P. (2005). Fantastic topographies: Neoliberal responses to Aboriginal land claims in British Columbia. *Canadian Geographer / Géographe canadien, 49*(4), 352–66. http://dx.doi.org/10.1111/j.0008-3658.2005.00101.x

Sayer, A. (2005). *The moral significance of class*. Cambridge: Cambridge University Press. http://dx.doi.org/10.1017/CBO9780511488863

Shamir, R. (2008). The age of responsibilization: on market-embedded morality. *Economy and Society, 37*(1), 1–19. http://dx.doi.org/10.1080/03085140701760833

Stevenson, M.L. (2001). Visions of certainty: Challenging assumptions. In *Speaking truth to power: A treaty forum* (113–33). Ottawa: Law Commission of Canada.

Tennant, P. (1996). Aboriginal peoples and Aboriginal title in British Columbia politics. In R.K. Carty (Ed.), *Politics, policy and government in British Columbia* (45–64). Vancouver: UBC Press.

Thompson, W.-A. (1999, 19 April). Everything's up for grabs. *BC Report*, 24–5.

Tully, J. (2001). Reconsidering the B.C. treaty process. In *Speaking truth to power: A treaty forum* (3–17). Ottawa: Law Commission of Canada.

Turner, R. (2008). *Neoliberal ideology: History, concepts and policies*. Edinburgh: Edinburgh University Press.

Warner, M. (2002). *Publics and counterpublics*. New York: Zone Books.

Woolford, A. (2005). *Between justice and certainty: Treaty making in British Columbia*. Vancouver: UBC Press.

PART 3

Neoliberal Cities? Police and Ad Hoc Governance

8 Urban Neoliberalism, Police, and the Governance of Condo Life

RANDY K. LIPPERT

About contemporary urban governance there is a familiar story in which neoliberalism plays a starring role. Specifically, it is said that new strategies, technologies, and authorities have emerged on the global urban stage to reshape and manage urban life and space in new directions (Brenner & Theodore, 2002; Isin, 1998; Keil, 2009; Kern, 2010; Lippert, 2007; Ruppert, 2006). This neoliberal governance, or "political-economic governance premised on the extension of market relationships" (Larner, 2000: 5), is especially evident within ambitious efforts to gentrify or, less honestly, to "regenerate" urban economies and spaces (see Blomley, 2004: 31). Typically, this means a stark reduction of municipal expenditures for services and social housing for urban populations, deregulation, a move towards entrepreneurial planning, and a valorization of the private sector's alleged capacity to deliver services and housing more efficiently than the state. One form closely associated with these urban neoliberal projects is the condominium corporation (hereafter condo).

Examining condos as a form of urban governance via ethnographic research that combines personal interviews, observation, and collection of text-based materials, this chapter seeks to refine understandings of the governmental elements that comprise and shape urban governance. Condos possess attributes consistent with urban neoliberalism. But largely absent from governmentality-related accounts of urban neoliberalism is the acknowledgment of other relevant governing logics, including those operating in these and other private (or privatizing) urban realms, or across private-public boundaries. One of these logics Michel Foucault called "police" (Foucault, 2007), which attends to the minutiae of urban life, or as Foucault aptly suggested, to "little things."

This chapter argues in relation to condos that the full significance of "police" in contemporary urban governance has gone unrecognized and that ethnography, broadly conceived, may be one vital means of exploring "police."

In condos "police" is found to initially coerce unit owners, as well as subsequently target myriad aspects of their conduct and other "little things" that constitute everyday urban life. "Police" practices are found increasingly to stem from various private agents' activities in the governance of condos and are becoming more and more marketized. Yet these practices still cannot be fully accounted for by reference to urban neoliberalism alone. All this underscores the need for more research into these private urban arrangements and practices using ethnography. Unlike research based entirely on publically available documents, ethnography can help discern how neoliberalism relates to other logics, including "police," particularly in growing private realms that are notoriously difficult to access or are hidden from view. Analysts can use ethnography to explore what some would consider mundane private realms to discover the little rules that constitute the urban at a distance from city councils where neoliberalism articulates with "police." Such an exploration matters because it can help expose regressive effects of combinations of such logics and allow thinking about alternatives.

The remainder of this chapter unfolds in four parts. The first section considers urban neoliberalism and "police" in recent governmentality-related literature with an eye to identifying the specific contribution that ethnographic research can make. The second section describes the emergence and characteristics of condos in order to raise doubts about their characterization as neoliberal. In this section the value of using ethnography to discover and explore "police" in condos and its relations with neoliberalism are discussed. Building on these sections, surveillance and rules in condos are discussed next. Finally, the free-rider problem (see Olson, 1965) and the marketization of "police" in relation to condos and neoliberalism is taken up, which shows how they are increasingly intertwined. Some concluding remarks follow.

Urban Neoliberalism, Police, and Governmentality

Governmentality studies and related work from across the disciplines, including political science, geography, and sociology, have documented complex shifts in the governance of western cities in the past three decades, often using "urban neoliberalism" or "neoliberalization" to describe these changes (see Brenner & Theodore, 2002; Hackworth, 2006; Kern, 2010). For example, they speak to how and/or why private forms of urban governance have arisen to replace or supplement rolled-back municipal bureaucracies and public services overseen by city councils, and they reveal how market-based urban housing schemes have emerged while urban public housing and rent-control programs have been reduced or altogether eliminated (Blomley, 2004).

There is a tendency in the literature to deploy a total or master narrative of urban neoliberalism and to suggest that this logic accounts for the contours of contemporary urban governance assemblages, but such an approach is empirically doubtful and somewhat inconsistent with the original thrust of governmentality studies inspired by a "later Foucault" (see O'Malley, 2001; Rose, O'Malley, & Valverde, 2006; Walters, 2012). Recent work on urban neoliberalization from within the governmentality literature and from structural perspectives occasionally recognizes resistances, holes, or anomalies in this otherwise totalizing configuration (e.g., Wilson, 2004). However, the aspects that are recognized to be outside or alongside are not typically characterized as eliciting a particular logic or rationality. They are defined only by what they are not. Thus, one encounters, among others, "contingent urban neoliberalism" (ibid.), "variegated neoliberalization" (Brenner, Peck, & Theodore, 2009), and "roll-with-it neoliberalization" (Keil, 2009). While they are well intentioned and often adroit, the awkwardness of these newer concepts nonetheless underscores the fact that urban neoliberalization is notoriously incomplete when studied in specific sites, and it highlights the related failure to specify the nature of what operates outside or alongside these processes in cities. Put another way, we might ask: what are the *contingencies* that *do not* lead to urban neoliberalism; what lies *between* brightly coloured neoliberal spots in cities; and with *what*, exactly, must neoliberal governance *roll* in cities?

Perhaps more satisfying approaches might be found by using additional orientating concepts such as "governing from below" (see O'Malley, 1996) that potentially recognize logics, technologies, and authorities beyond, adjacent to, or below neoliberalism. These elements tend to be neglected, partially owing to widely perceived banality, especially evident in the case of condos, but also to an assumed immorality and criminality in instances of urban crime networks (Lippert & Stenson, 2007), or to invisibility where representative texts and detailed empirical inquiry (including ethnography) are absent. Yet it is these elements that partially comprise novel urban assemblages of rule.

This concept of urban "assemblage" itself, well known in governmentality scholarship (see also Li and Valverde in this volume), "allows and encourages the study of the heterogeneous connections between objects, spaces, materials, machines, bodies, subjectivities, symbols ... that 'assemble' the city in multiple ways" (Farias, 2010: 14). Thus, the urban assemblage concept invites inquiry into logics besides neoliberalism in the condo form.

Collier's (2009, 2012) recent work on governmentality and neoliberalism is pertinent here. He writes that neoliberalism in structural approaches is sometimes depicted as the big leviathan: "the entire ensemble of elements is identified with neoliberalism ... [and] neoliberalism grows bigger, and becomes more fundamental, more structural and structuring, than other things in the field" (2012: 189). He suggests that what "is now required [instead is] to show how styles of analysis, techniques or forms of reasoning associated with 'advanced liberal' government are being recombined with other forms, and to diagnose the governmental ensembles that emerge from these *recombinations*" (2009: 99; emphasis added). The following analysis seeks to accomplish this goal, particularly in relation to the "free-rider problem" and "police."

"Police" is a neglected but pertinent logic highlighted in Foucault's writings (1988). Of special relevance are Foucault's statements about an early tether between "police" and the urban. Thus, Foucault remarked in his famous *Security, Territory, Population* lectures when reflecting on earlier writers, including Domat from the eighteenth century, that "to police and to urbanize is the same thing" (2007: 337). These increasingly quoted Foucauldian reflections are exquisitely suggestive for understanding contemporary urban governance assemblages. In his lectures Foucault quotes Catherine II on a code of police: "The things of police are things of each moment ... Police is concerned with little things" (ibid.: 340). "Police" refers to a regulatory power, sometimes called "the police power," entailing coercive measures to ensure a community's welfare (see Dubber & Valverde, 2006; Novak, 1996: 70). "Police" is the power to govern humans and things, to treat the city, or a small portion thereof like a household, and in this sense can be seen as patriarchal (Dubber, 2004). The logic of "police" dreams of regulating everything urban and targets things and their positioning as much as human conduct. The objects of "police" are things of the urban (see Foucault, 2007: 336).

When invoked in governmentality studies, "police" typically appears as the defunct antecedent of early liberalism (e.g., Hunt & Wickham, 1994), thus serving as a convenient foil for Foucault's account of liberal governmentality (see Walters, 2012: 28). But there is little evidence "police" or the police power has become less prevalent in western cities through the twentieth century or since urban neoliberalization commenced. "Police" has not yet been fully explored and is not accounted for by reference to neoliberalism alone. Leading governmentality analysts in socio-legal studies such as Hunt (2006), Blomley (2011), and Valverde (2011) also have begun to recognize "police" in the urban. Their groundbreaking work points the way and can be extended. Hunt (2006), for example, has explored, as a form of "police," the emergence of regulations and signals in cities to confront that distinctively urban problem called "traffic," which, of course, continues to proliferate. Blomley (2011) has cogently argued that pedestrianism is a form of "police" as well. He argues that pedestrianism avoids governing through rights and instead elicits "an attention to placement and flow" (ibid.: 4). Finally, Valverde (2011) has asserted that premodern and modern rationalities coexist in cities. This is a vital point. In urban governance there is a "pragmatic approach that uses both old and new gazes, pre-modern and modern knowledge formats, in a nonzero-sum manner and in unpredictable and shifting combinations" (ibid.: 281). Thus, here there is emphasis on combination rather than a singular totalizing process. Nonetheless, scholars have so far tended to overlook private eyes among the "old and new gazes." Valverde notes that to fully understand urban governance there is a need to study these private realms, too: "Public laws and rules did not have a monopoly on regulation, because such private actors as insurance companies also imposed rules ... lenders as well as realtors also imposed their own private but nevertheless compelling regulations ... A full genealogy of urban governance would have to include the myriad private ... regulatory structures that converged on different kinds of property owners and different kinds of properties" (ibid.: 292).

This assertion can be extended. It is regulation by insurance firms *of* property owners, but also regulation *by* condo boards and other property owners concerning the "little things" that form private property that are essential to understanding urban governance assemblages.

In the work of governmentality scholars noted above, "police" flows mostly from municipal government. But, as will be shown below, contemporary "police" need not be so tightly tethered to the municipal level. The preceding work drawing on Foucault's notion of "police" is a compelling, suggestive start of a move towards more nuanced accounts of contemporary urban governance assemblages. However, private arrangements that target behaviour in residential and other urban spaces ought to be explored too. Valverde (2011: 287) recently criticized Novak's work on the police power because it "remains almost wholly confined to the municipal level" when, in fact, it is easily recognized at the federal level. Yet governmentality analysts also can look *lower* for instances of governing from below (including in the workings of condos), where placement and flow are paramount concerns. How "police" *combines* with other logics, including neoliberalism, in these lower realms remains unexplored.

Ethnographic research involving personal interviews, observation, and collection of private text-based materials can effectively explore "police" and "little urban things" in these lower private urban realms. Certainly, ethnography has been advocated before to study neoliberalism (Ferguson & Gupta, 2002), but it is not necessarily useful to look *solely* for the hidden neoliberal character of urban governance assemblages on the ground. The private realms discussed below are ultimately subject to the courts. However, in practice, as discovered through interviews with condo board members, comparatively few "little things" in condos that could be litigated actually enter the courts (see Lippert, 2012). Rather, there are more complicated governing arrangements in place within condos that cannot be reduced to the master narrative of neoliberalism, and through ethnography they can be revealed. It is precisely the introduction of common elements and spaces in these "lower" condos, beyond individual ownership, that either demands "police" or permits "police" to peacefully coexist with neoliberal demands that urban private property be protected and maintained. It is the creation of common elements and spaces that tethers the condo so closely to the urban, since *by definition* urban space is at a premium. These spaces are carved out with tiny rules that are often necessarily broken. The urban is where we can expect to find "police," and it is perhaps best uncovered via ethnographic methods.

Condo Context

The condo is an overlooked form of contemporary urban governance. This form entails governance of residential urban spaces and has expanded rapidly since the 1960s. Owning private property in this urban space is sufficient for membership, and the scope of governance stemming from these spatialized urban clubs is at times extensive. It includes garbage collection, transit, physical security, and recreational services for condo members. The condo is a private "spatial enclosure" that leads to social and spatial exclusion in part through a bias towards property-owning members and against others (especially renters and visitors in condos) who do not directly pay fees or serve on boards). These forms of exclusion are consistent with effects widely attributed to urban neoliberalism. Indeed, condos are commonly associated with neoliberalism (e.g., Kern, 2010). The aim of this section is not to explore the condo form in exhaustive detail but rather to consider some features that raise doubt about the full appropriateness of labelling them "neoliberal," and to suggest the presence of "police" in the governance of condos and its corresponding attention to "little things." Recent work, mostly by urban geographers, is beginning to explore condos and "condo-ization" with a critical eye on using historical methods and surveys (e.g., Lasner, 2012; Rosen & Walks, 2013), but rarely are they studied, directly or obliquely, using ethnographic methods broadly defined (but see Kern, 2010; Lippert, 2012; Lippert & Steckle, 2014). In what follows I draw from an ethnographic project conducted over a ten-year period. Data sources included numerous personal interviews with condo owners, property managers, and board members as well as visits to condo buildings in Toronto (Canada) and New York City (United States), where observational field notes were taken and text-based materials were collected.

Condos are market-driven housing schemes that often displace or replace social, public, or subsidized housing and in this sense are neoliberal in orientation. But the theoretical relevance of these arrangements does not stop at point of purchase (Lippert & Steckle, 2014); it is how condos are managed afterward that is arguably more interesting and relevant to understanding contemporary urban governance assemblages.

If "police" is patriarchal in treating the object of government as a household (see Dubber, 2004), it is perhaps not inconsequential that Keith Romney, who helped draft the first continental US legislation in Utah in 1960 and was the developer's lawyer, is called the "father of condominiums." His first offspring was a condo created from a planned "cooperative" called "Graystone Manor" in Salt Lake City. The condo is thus an invented form combining the common ownership of "co-ops" with traditional individual (detached) home ownership. In 1961 the US Federal Housing Authority (FHA) used a 1958 Act from Puerto Rico to draft model condo legislation, later used in many US states. Puerto Rico had earlier encountered a shortage of housing, high housing costs, and land scarcity. The first two of these conditions were occurring in some US cities (see Nelson, 2005). Since the FHA's drafting of model condo legislation to solve these distinctively urban problems and the construction of Graystone Manor in the following year (Romney, 1974), condos have proliferated across North America at a remarkable rate. Condo legislation had formed in all states by 1967. Condo law emerged in Canada in the late 1960s and corresponding housing schemes have grown rapidly since, particularly in Vancouver and Toronto (see Harris, 2011; Kern, 2010).

The recognition of a "free-rider problem" in relation to the shift from cooperative to condo cohered with assignment of risk in residential living arrangements to individuals. Yet the commonly owned elements needed ongoing maintenance. Residents with a propensity to consume non-excludable resources were imagined drawing common resources at individual owners' expense. Condo law anticipated the free rider, and its primary solution was a regular mandatory fee levied on all unit owners to maintain the condo's common elements and spaces. However, as condos flourished, this notion of the free rider became a way of conceiving of many governance problems. The "free-rider problem" remained salient across a spectrum of governance issues (see Nelson, 2005: 116–18), including, as noted below, the issue of unit-owner participation on condo boards. Condo law also introduced measures and procedures to ensure unit owners' compliance with by-laws and rules and a democratically elected board drawn from the membership to oversee these arrangements (see also Risk, 1968). Owing to the wide range of services these boards provide and oversee and the size of their budgets (with the development of larger and larger condo buildings, often in the millions of dollars), condo corporations are often termed a "fourth level of government" (e.g., Silverstein, 1992).

Condo Rules and Surveillance

Exploring private condo living arrangements via ethnography reveals that rules are intensive; they specify whether and how you can make noise after 10 p.m.; the size of your cat; what you can wear in recreation areas; how long you can leave your vehicle in front of the building; how many guests, and for how long, you can entertain (so as not to create free riders); and whether and how you can remodel your kitchen or paint your exterior door. Many of these rules are about aesthetics but they also deal with positioning and flow through space, and the timing thereof, as myriad examples from resident informants from condos reveal: "There are rule infractions ... some people want to renovate their apartment or improve it and don't follow the rules about when they are going to do it or ... having construction at odd hours ... So ... there's an effort by the management to ... discuss these issues with them and educate them as to what the rules are ... about which hours would be suitable ... charges for moving in and moving out, and things like that. All of those are established as needed to address ... inappropriate behavior" (Condo board member, Interview 10). Another resident indicated there were "people who will leave their bicycles where the trash compactor is; instead of taking it downstairs and putting it away, they'll leave it there. I'm kind of like, 'Why would you do that, you have an elevator?' So ... you know you have things like that ... there used to be incidents where dogs were peeing in the hallways so they implemented a bylaw where if your dog pees in the hallway ... you're going to get fined" (Condo resident, Interview 17). In another condo, "each floor has a garbage chute disposal area; sometimes people leave boxes on the floor when the sign says specifically 'Do not leave boxes on the floor,' or some people will use the area after 10 p.m. at night when there's a sign there that says 'Don't use the garbage after ten'" (Condo resident, Interview 65).

Besides aesthetics, some rules pertain more specifically to flow through condo spaces: "Our halls are small because there're only three units on a floor, so we are not to put anything out in the hall. No rugs, nothing. I mean, you put it in your unit; you don't put your boots out in the hall; you don't put your strollers in the hall" (Condo resident, Interview 38).

Often surveillance is demanded by "police" (see also Lippert & Murakami-Wood, 2012):

[When] I was on the board I had that kind of mentality ... my nickname was the Building Sheriff because I would sort of try to keep people to the rules. You know, like ... especially if I hear the dog; [when] the dog first moved into the building, I was trying to figure out where the dog was barking from. I went out of my apartment and I went down and I was listening for the noise, and I saw which apartment it was coming from and I sort of knew, confirmed my suspicions ... I wish I had a camera locked in the elevator to catch the dog urinating in the elevator. (Condo resident, Interview 18)

Sometimes municipal by-laws such as those pertaining to fire codes that also reflect "police" penetrate condo life and encourage further surveillance at the condo level: "Take the stroller, put it in the fire stairs, because there's no place else for it. You can't do that and what happens is the fire marshal comes by on a regular basis and walks the stairs, and if there's anything in the stairs the building gets fined, and the buildings [have] been fined more than once. Now the Super, on a regular basis, kind of patrols so that we're not in violation of any of those fire codes" (Condo resident, Interview 18).

In many condos observed during ethnographic research, there was actually an array of security cameras installed throughout condo spaces, as residents indicated, to record myriad rule infractions: "We have cameras in the elevators and then we have cameras all around the building and in the common elements, but I don't have any cameras on the floors, so if somebody decides to leave their pizza boxes outside the elevator door I don't know who's done it ... [so we have to] check the garbage chute rooms on every floor to see what some idiot has dumped on the floor" (Condo resident, Interview 2). In other condos, there were "kids on the roof throwing rocks off the roof ... But you know [what] most people are like, if you approach them nicely and civilly and say 'Excuse me, you know we got some videotape that shows that your kids were on the roof throwing some rocks off. I know you don't think that's a good idea. Can you make sure that doesn't happen again?'" (Condo resident, Interview 15).

Interviews with condo informants revealed a preoccupation with and resistance to a dizzying number of rules that sought to control all manner of "little things." As another owner aptly remarked, her condo board decides rules on the "granular level": "The board meets monthly. I know that they tend to discuss issues that are on a pretty granular level. Our board is quite hands-on and if ... in doubt they tend to issue a rule ... They tend to be pretty heavy-handed with rules and signs everywhere" (Condo resident, Interview 4).

Surveillance consistent with "police" also activates human agents in condos, sometimes as property managers, security patrols, and door-persons or concierges, who become aware of rule infractions concerning "little things," such as admitting visitors to recreational areas in the building late at night, owing to their temporal and spatial positioning at points of access and egress. Unlike the notice given regarding state-sponsored surveillance in public spaces, as limited and inconsistent a practice as it may be where liberal subjects are concerned (see Lippert, 2009), there is typically little or no effort to alert condo residents of their rights regarding practices conducted by these various agents.

As implied by the accounts above, there are also those called "strata Nazis" or "condo commandos" in condo governance: owners who voluntarily patrol and collect intelligence about other owners' conduct and about the changing condition of common elements and spaces. These agents are bent on enforcing rule infractions as board members or they demand compliance from condo boards. While the extent of their role in relation to the other "police" agents above remains uncertain, their very existence is often kept hidden to avoid dissuading others from wanting to purchase their way into the "condo club." While examining internal or private texts regarding condo governance would illuminate relevant governmental discourses if they could be acquired, it is difficult to imagine understanding the resulting network of agents and corresponding rule enforcement and avoidance that sometimes involves covert practices, known only to insiders, without ethnographic methods.

Free-Rider Problems and "Police" Marketization

Free-Rider Problems

The foregoing questioned whether condos are wholly neoliberal. If condos are increasingly "fourth levels of governments," they may well be understood better by reference to "police." This is not surprising, since, while this form has neoliberal features, it is known to yield regressive and exclusionary effects typically attributed to urban neoliberalism; it also predates the arrival of neoliberal governance in earnest. While condos appeared in the 1960s in North America, neoliberal rationalities began to shape governmental programs only in the 1970s (Harvey, 2005: 2). For condos, what was novel at their inception was not any obviously neoliberal features; for example, these private, spatialized clubs were not initially accompanied by a valorization of private actors' capacity to efficiently deliver urban services and housing; rather, central to initial condo arrangements, as outlined in enabling legislation, was a perceived need to coerce free riders. To be sure, these imagined free riders are subjects deemed to pursue narrow economic incentives (Bartholomew & Hunt, 1990: 54), and to that extent they are a form of neoliberalism's "economic man" (see Li in this volume). As well, condo legislation and related governing arrangements are consistent with neoliberal rationalities to the extent that they seek to establish and help guarantee a form of private property (see, e.g., Harvey, 2005: 2). However, condos as forms of urban governance go beyond merely guaranteeing private property and are more than merely neoliberal technologies. Thus, one notes there is little concern in the rationality of "police" as enacted in condo governance about free riders' motivations, such as whether they are responsible or rational, as one might expect with governance arrangements informed by neoliberal rationalities. Rather, the concern is simply that free riders are overloading the train at the expense of the welfare of other travellers. The motivation of those boarding or stepping off is not deemed important. That the riders on the rails are reckless risk takers or more responsible agents is largely irrelevant, and therefore there is no attempt to act on these actors' dispositions. Thus, in condo arrangements there is a discernible "police" that can articulate with but cannot be reduced to urban neoliberalism. Exploring the former also can illuminate the latter, as is shown in the next subsection.

Marketization of "Police" Practices

As initially conceived in legislation, in condos "police" practices would stem from boards only. There was no anticipation these practices would be later marketized through private insurance, property management, and private security. Private insurance (see Ericson, Doyle, & Barry, 2003) benefits from increased reliance on privatized (rather than state or public) forms of risk avoidance. Since the 1960s more risks associated with the governance and workings of private condos have been identified, and these risks have been increasingly packaged as insurance commodities to be bought and sold (see Lippert, 2012). As this change has occurred, there is a sense in which insurance has taken over some control from boards of related "police" practices. For example, board directors' insurance, which has been increasingly marketed as the capacity of board decision making to effectively manage condo life without major errors, has been more and more called into doubt. This doubt includes whether boards are ensuring that there is enough money in a building's reserve fund – consistent with condo legislation of the time – for later replacement and repairs of a building's systems and infrastructure. Such a concern is already evident in formal reviews of condo legislation conducted in the late 1970s (see Kealey, 1977). There has also been increasing concern about the risk of boards' not following numerous other applicable state laws, ranging from ensuring that the premises are safe for occupiers (see Occupiers' Liability Act, 1990) to ensuring that information about unit holders is collected and stored in ways that are consistent with information privacy legislation. Interviews revealed that board insurance is increasingly purchased to backstop any such omissions or errors among board members. Through threat of policy cancellation or rate increases, these insurance companies in effect dictate to board members that certain safety-related practices and rules must be followed. The influence of insurance on marketizing "police" practices has rapidly expanded and intensified in the condo world. Similarly, the property management industry can be observed working in conjunction with the dictates of private insurance as a form of marketization of "police." Property management is rapidly expanding as condo life and, specifically, solutions to free-rider problems are increasingly commodified (see Lippert & Steckle, 2014). There is a sense in which *condo* property management in particular is premised on a "free-rider problem" (see also Chen & Webster, 2005; Nelson, 2005). Interviews consistently revealed that unit owners are reluctant to participate on boards or to volunteer services in their own condo buildings, preferring instead to let others manage condo life on their behalf. Property management's entry into the condo world presumes this problem will be ongoing, but, as noted above, this issue was unanticipated in the original condo legislation that assumed all owners would reside in and willingly manage the intimate affairs of condo living by periodically serving as board members. Finally, condos increasingly contract private security firms to carry out "police" practices and provide concierges or doorpersons to monitor access and egress and security arrangements.

Governance of the myriad of little issues that arise in condos is therefore increasingly provided through these various private institutions, despite the fact that only boards have statutory legal authority to govern condo affairs. The free-rider problem and the rationality of the "police" are becoming more closely tethered to the goals of commercial institutions implicated in these increasingly marketized arrangements. However, despite this situation, we can best make sense of current urban governance arrangements by viewing them as informed by distinct but intertwined rationalities of "police" and neoliberalism rather than by viewing one rationality as being reducible to the other.

Concluding Remarks

Contrary to what is implied in the literature, condos are not obviously being "responsibilized" by carrying out municipal government instructions "at a distance" within defined spaces (see Rose, 1999:

49–50). However, condos are operating in ways that are consistent with "police," including being concerned with aesthetics and the positioning of objects and persons in and the flow through their defined spaces and the timing thereof. Condos' power to effectively extract mandatory resources from members, make and unmake rules of conduct and positioning, provide or withdraw physical security, and include or exclude persons in relation to these new urban realms is significant. The rationality of "police" is therefore not merely an unforeseen obstacle or merely something with which neoliberalism must roll or roll over. Fully understanding contemporary urban governance assemblages therefore requires a study of the rationality of "police" too.

The tight connection to property markets has resulted in condo development being associated with neoliberalism (see Kern, 2010). However, a close examination of the governance of condos reveals that their governance is informed by neoliberal and non-liberal rationalities and thus involves combining new elements with elements already in place rather than relying on novel invention. If governmentality is about inventiveness, it is more about using what is available and less about deploying technologies fresh from "eureka" moments. The foregoing reveals one way neoliberalism can sustain itself or at least avoid conflict. Contemporary governance of condos seems to be informed by multiple rationalities, where non-liberal rationalities remain in place after the arrival of neoliberal rationalities. This chapter underscores the need for more research into private arrangements and elements that comprise urban governance assemblages, using ethnography as one key method. However, it also implies that ethnography should not be deployed to the neglect of exploring the historical trajectory of these elements that constitute such assemblages.

To explore condos and "police" is not to valorize them. Especially because "police" is claimed to be patriarchal, ostensibly uninterested in subjects' motivations, and in practice seems to work largely without external legal or other scrutiny, there is reason to doubt its potential role in progressive urban governance (but see Valverde, 2011: 279). As Foucault (2007: 339) starkly states, "Police is not justice." This matters because there is no shortage of social and spatial exclusion and other regressive practices in this new urban world, and it is therefore essential to know how and which specific logics lead to these effects before conceiving of alternatives. Foucault's later work provided socio-legal and other kinds of scholars with a relevant legacy of concepts like "police" to use. Given that he wrote relatively little about how neoliberalism

shapes particular policy domains as opposed to how key thinkers articulated this rationality, it is perhaps surprising how much governmentality work has invoked neoliberalism while neglecting "police" and
other logics, alone and in combination. For now it remains uncertain
whether "police" merely plays a supporting role for urban neoliberalism's popular performance or whether "police" remains the venerable
star of urban governance at municipal and lower levels. Neoliberalism
may be but a newcomer demanding revisions to an old urban script,
at least to the extent that "police" function in condos is becoming marketized. Examining condos reveals that the rationalities of "police" and
neoliberalism coexist and articulate in the governance of urban residential life. But the form that such police-neoliberal configurations will
take cannot be identified in advance. They must be studied in context,
where myriad little urban things become serious concerns, and, if the
study of condo governance and life is illustrative, ethnography ought
to be embraced as an enlightening means of doing so.

ACKNOWLEDGMENT

Parts of this chapter were first published in a 2014 issue of *Foucault Studies*, *18*, 49–65.

REFERENCES

Bartholomew, A., & Hunt, A. (1990). What's wrong with rights? *Journal of Law and Inequality, 9*(1), 1–58.

Blomley, N. (2004). *Unsettling the city: Urban land and the politics of property.* New York: Routledge.

Blomley, N. (2011). *Rights of passage: Sidewalks and the regulation of public flow.* New York: Routledge.

Brenner, N., & Theodore, N. (2002). Cities and geographies of "actually existing neoliberalism." *Antipode, 34*(3), 349–79. http://dx.doi.org/10.1111/1467-8330.00246

Brenner, N., Peck, J., & Theodore, N. (2009). Variegated neoliberalizations: Geographies, modalities, pathways. *Global Networks, 10*(2), 188–222.

Chen, S.C.Y., & Webster, C.J.C. (2005). Homeowners associations, collective action and the costs of private governance. *Housing Studies, 20*(2), 205–20. http://dx.doi.org/10.1080/026730303042000331736

Collier, S.J. (2009). Topologies of power: Foucault's analysis of political government beyond "governmentality." *Theory, Culture & Society, 26*(6), 78–108. http://dx.doi.org/10.1177/0263276409347694

Collier, S.J. (2012). Neoliberalism as big leviathan, or …? A response to Wacquant and Hilgers. *Social Anthropology, 20*(2), 186–95. http://dx.doi.org/10.1111/j.1469-8676.2012.00195.x

Dubber, M.D. (2004). "The power to govern men and things": Patriarchal origins of the police power in American law. *Buffalo Law Review, 52*(4), 101–66.

Dubber, M.D., & Valverde, M. (Eds). (2006) *Police and the liberal state.* Stanford: Stanford University Press (Stanford Law Books).

Ericson, R.V., Doyle, A., & Barry, D. (2003). *Insurance as governance.* Toronto: University of Toronto Press.

Farias, I. (2010). Introduction: Decentering the object of urban studies. In I. Farias & T. Bender (Eds), *Urban assemblages: How actor network theory changes urban studies* (1–24). New York: Routledge.

Ferguson, J., & Gupta, A. (2002). Spatializing states: Toward an ethnography of neoliberal governmentality. *American Ethnologist, 29*(4), 981–1002. http://dx.doi.org/10.1525/ae.2002.29.4.981

Foucault, M. (1988). Politics and reason. In L. Kritzman (Ed.), *Politics, philosophy, culture, interviews and other writings, 1977–1984* (57–85). New York: Routledge.

Foucault, M. (2007). *Security, territory, population: Lectures at the Collège de France, 1977–1978.* New York: Palgrave Macmillan. http://dx.doi.org/10.1057/9780230245075

Hackworth, J.R. (2006). *The neoliberal city: Governance, ideology and development of American urbanism.* Ithaca: Cornell University Press.

Harris, D.C. (2011). Condominium and the city: The rise of property in Vancouver. *Law & Social Inquiry, 36*(3), 694–726. http://dx.doi.org/10.1111/j.1747-4469.2011.01247.x

Harvey, D. (2005). *A brief history of neoliberalism.* New York: Oxford University Press.

Hunt, A. (2006). Police and the regulation of traffic: Policing as a civilizing process? In M. Dubber & M. Valverde (Eds), *The new police science: The police power in domestic and international governance* (168–84). Stanford: Stanford University Press.

Hunt, A., & Wickham, G. (1994). *Foucault and law: Toward a sociology of law as governance.* Boulder, CO: Pluto Press.

Isin, E. (1998). Governing Toronto without government: Liberalism and neoliberalism. *Studies in Political Economy, 56,* 169–91.

Kealey, D., (1977) *Report of the Ontario residential condominium study group*. Ministry of Consumer and Commercial Relations. Toronto: Study Group.

Keil, R. (2009). The urban politics of roll-with-it neoliberalization. *City*, 13(2–3), 231–45.

Kern, L. (2010). *Sex and the revitalized city: Gender, condominium development and urban citizenship*. Vancouver: UBC Press.

Larner, W. (2000). Neoliberalism: Policy, ideology, governmentality. *Studies in Political Economy*, 63, 5–25.

Lasner, M.G. (2012). *High life: Condo living in the suburban century*. New Haven: Yale University Press.

Lippert, R.K. (2007). Urban revitalization, security, and knowledge transfer: The case of broken windows and kiddie bars. *Canadian Journal of Law and Society*, 22(2), 29–53. http://dx.doi.org/10.1017/S0829320100009340

Lippert, R.K. (2009). Signs of the surveillant assemblage: Privacy regulation, urban CCTV, and governmentality. *Social & Legal Studies*, 18(4), 505–22. http://dx.doi.org/10.1177/0964663909345096

Lippert, R.K. (2012). Governing condominiums and renters with legal knowledge flows and external institutions. *Law & Policy*, 34(3), 263–90. http://dx.doi.org/10.1111/j.1467-9930.2012.00364.x

Lippert, R.K., & Murakami-Wood, D. (2012). The new urban surveillance: Technology, mobility, and diversity in 21st century cities. *Surveillance & Society*, 9(3), 257–62.

Lippert, R.K., & Steckle, R. (2014) Conquering condos from within: Condoization as urban governance and knowledge. *Urban Studies*. Available at http://usj.sagepub.com/content/early/2014/12/19/0042098014562332.abstract. Accessed 1 October 2015.

Lippert, R.K., & Stenson, K. (2007). Urban governance and legality from below. *Canadian Journal of Law and Society*, 22(2), 1–8. http://dx.doi.org/10.1017/S0829320100009315

Nelson, R.H. (2005). *Private neighborhoods and the transformation of local government*. Washington, DC: Urban Institute Press.

Novak, W.J. (1996). *The people's welfare: Law and regulation in nineteenth century America*. Chapel Hill: University of North Carolina Press.

Occupiers' Liability Act, R.S.O. 1990, c. O.2. (1990). Available at http://www.ontario.ca/laws/statute/90o02. Accessed 7 March 2016.

Olson, M. (1965). *The logic of collective action*. Cambridge, MA: Harvard University Press.

O'Malley, P. (1996). Indigenous governance. *Economy and Society*, 25(3), 310–26. http://dx.doi.org/10.1080/03085149600000017

O'Malley, P. (2001). Genealogy, systematization and resistance in "advanced liberalism.". In G. Wickham & G. Pavlich (Eds), *Rethinking law, society and governance: Foucault's bequest* (13–25). Portland, OR: Hart.

Risk, R.C.B. (1968). Condominiums and Canada. *University of Toronto Law Journal, 18*(1), 1–72. http://dx.doi.org/10.2307/825167

Romney, K.B. (1974). *Condominium development guide: Procedures, analysis, forms.* Boston: Warren, Gorham & Lamont.

Rose, N. (1999). *Powers of freedom: Reframing political thought.* Cambridge: Cambridge University Press. http://dx.doi.org/10.1017/CBO9780511488856

Rose, N., O'Malley, P., & Valverde, M. (2006). Governmentality. *Annual Review of Law and Social Science, 2*(1), 83–104. http://dx.doi.org/10.1146/annurev.lawsocsci.2.081805.105900

Rosen, G., & Walks, A. (2013). Rising cities: Condominium development and the private transformation of the metropolis. *Geoforum, 49*(1), 160–72. http://dx.doi.org/10.1016/j.geoforum.2013.06.010

Ruppert, E. (2006). *The moral economy of cities: Shaping good citizens.* Toronto: University of Toronto Press.

Silverstein, A. (1992, 10 October) First fluffy, now Old Mickey, are booted out. *Toronto Star*, E1.

Valverde, M. (2011). Seeing like a city: The dialectic of modern and pre-modern ways of seeing in urban governance. *Law & Society Review, 45*(2), 277–312. http://dx.doi.org/10.1111/j.1540-5893.2011.00441.x

Walters, W. (2012). *Governmentality: Critical encounters.* London, New York: Routledge.

Wilson, D. (2004). Toward a contingent urban neo-liberalism. *Urban Geography, 25*(8), 771–83. http://dx.doi.org/10.2747/0272-3638.25.8.771

9 Ad Hoc Governance: Public Authorities and North American Local Infrastructure in Historical Perspective

MARIANA VALVERDE

Today's public-private partnerships (P3s) are generally thought of as creatures of neoliberal governance trends. It is undoubtedly true that the political effects of many if not most of these partnerships and looser networked arrangements are often in keeping with neoliberal political and economic aims. However, this paper suggests that today's local partnerships are being built with legal tools and governance forms that draw on a long history of public authorities, "special-district governments," and other hybrid legal forms created, often in a purely ad hoc manner, to further locally based public purposes – entities that cannot be understood through the notions of "privatization" and "neoliberalism." Whether using ancient common-law forms such as the trust or the corporation, or going under the American label of "special district government," the fact is that for centuries local development, local transportation, and local prosperity have been promoted by entities that serve public purposes and often use public funds (though often alongside other funds, e.g., fees), but are not part of a ministry or a local government department and may behave like private corporations but are neither for-profit firms nor third-sector non-profits.

In other words: the genealogy of today's ubiquitous public-private partnerships is deeply rooted in a little-known history of local governance that cannot be reduced to unidirectional narratives (e.g., the rise of neoliberalism). Reviewing the literature on the history of local governance and special-purpose bodies suggests that any generalization about shifts from "public" to "private" are not justified by the evidence (see, e.g., Foster, 1997; Frieden & Sagalyn, 1989; Novak, 2001; Perry, 1995). Indeed, about the only generalization that is strongly supported by the available evidence is that, as entities from parks boards to public transit bodies to "turnpike trusts" to port authorities go on with their work, the tools with which they carry out their work have a remarkably ad hoc character. And, not surprisingly, the entities or loose networks of entities that have built major infrastructure projects in what is often a frankly ad hoc manner are better described by the term "assemblage" than by a term such as "institution" or even "partnership." The term "assemblage" signals an affiliation with the kind of post-institutional and even post-sociological theoretical and methodological moves found in actor-network studies and governmentality studies (for more on this topic see Li in this volume). By using the term "assemblage" I signal a desire to dynamically study the shifting, ever-changing artic-ulations of heterogeneous participants and interests that one finds in all governance, local or not; but I stress that my choice of terms such as "ad hoc" and "assemblage" is empirically rather than theoretically driven, in the sense that only after studying the relevant literature did I decide on the terminology (rather than starting out with the idea of studying assemblages). Focusing in as dynamic a manner as possible on assemblages has the added advantage of decentring both institu-tions (mainly, for local purposes, the municipal corporation) and the area of law (municipal law), the traditional objects of study of political science and legal scholarship, respectively.

The ad hoc manner in which legal and administrative and informal tools are used to bind "partners" together and carry out tasks such as building or operating infrastructure is especially marked in the United States and Canada. In the UK the central government has been able to flex its powers to impose sharp limits on the ability of local councils both to undertake work themselves and to partner with private sector bodies; in general, the English government has imposed a constrained form of standardization on local government since the days of the 1834 Poor Law (Loughlin, 1996). And today, many civil law countries, for example, in Latin America, have laws of general application governing public-private partnerships, an approach that, while not barring ad hoc entities similar to the thousands of public authorities that exist in the United States, imposes a certain amount of standardization on the legal and governance tools that can be used. But in the United States and in Canada local governance entities have rarely had contact with the cen-tral, federal state other than occasionally receiving discretionary grants or loans. For their part, provincial and state legislatures appear to have been much more concerned with defining and limiting the powers of municipal corporations (see, e.g., Frug, 1999; Levi &Valverde, 2006) than with governing (or even counting) the vast number of public and quasi-public special-purpose bodies that do a great deal of the unsung work of local governance.

Entities that combine public powers such as expropriation with the financial flexibility of private corporations continue to enjoy much autonomy while remaining largely invisible not only to the public but even to politicians and legislators, and they play a significant role in many of the infrastructure and urban development assemblages of which public-private partnerships are only one type. Thus, the reasons why entities that build and operate bridges and ports and that deliver local water, electricity, public transit, and other services should have developed in a markedly unplanned, ad hoc manner is a question of more than historical interest. That there is a world-wide mania for "partnered" governance and even "partnered government," in Larner and Butler's (2007) apt words is well known. However, to understand, concretely, how this worldwide phenomenon is currently unfolding, one needs to appreciate the very long history of local institutional creativity that has been the constant albeit relatively invisible supplement to the better-known stories about the mid-nineteenth-century separation of public from private corporations (Horwitz, 1982; Kaufman, 2008; Williams, 1985) and the slow but ineluctable rise of standardized municipal corporations subject to uniform rules (Frug, 1999; Loughlin, 1996).

Let us begin with a concrete illustration of the governance phenomena that my argument about "ad hoc governance" seeks to explain. In the 1930s and 1940s, the legislature of New York state created a range of different special-purpose authorities for the purpose of building bridges and other local infrastructures with newly available federal funds. Infrastructure funds were liberally provided by the Roosevelt administration during the New Deal, as is well known, but what is not so well known is that these funds were not made available to actual, legal cities, since the Roosevelt government shared the suspicions about urban politicians' corruption and wastefulness held by state legislators and courts (Edelstein, 1942–3; Elkind, 1997). The proliferation of special-purpose authorities in the New York City area during the New Deal underscores the fact that despite the Thatcherite enthusiasm for urban development corporations and "quangos," there is no necessary link between the form of the arms-length public or quasi-public local governance entity, on the one hand, and neoliberal political content on the other.

These not-quite-municipal entities, often endowed with public powers such as expropriation, are simultaneously also able to act like private corporations. One key development was the 1890s invention of the "revenue bond" – a financing tool that allowed local special-purpose public authorities, even if only slightly removed from city councils, to borrow substantial sums of money using novel forms of collateral, the debt acquired through novel financing then acquiring a critical political virtue: it did not count as part of the municipal debt. As early as 1898 the Supreme Court of the United States confirmed, in the Walla Walla case, that the bodies that came to be called "public authorities" or "special-purpose authorities" could use their future revenue as collateral to borrow money (Rose, 1953). By contrast, municipal corporations were (and often still are) restricted to issuing bonds secured by their current assets and were/are also usually subject to strict rules banning deficit budgets – rules that in the United States were and continue to be enshrined in state constitutions.

In Canada this novel tool was used when the newly created Toronto Harbour Commission borrowed large sums of money in 1912 to build a new port on a very large stretch of swampy ground, using as collateral the property that did not yet exist but would be created through landfill (Desfor & Laidley, 2011). The Toronto Harbour Commission example underlines a key virtue of the arms-length public authority: the commission did not, in fact, succeed in creating a wonderful new port to challenge Montreal's dominance in Great Lakes shipping, since, as is the case with many of today's P3s, their customer projections were wildly over-optimistic; but this failure had no direct effect on the municipal debt.

The (unsuccessful) Toronto Port plan is but one of myriad examples of governance innovation beyond (or beside) the municipal corporation. From the mid-nineteenth century onward there was a constant and often very pressing need for entities able to build, finance, and manage local infrastructure on a "supra-municipal," "inter-municipal," or simply non-municipal level. There was also a pressing need to find ways to finance these public works. History could have been different: if their fiscal and financing powers had grown rather than shrunk, municipalities (acting separately or together) could indeed have built and operated any number of infrastructures and transportation systems. But as it was, local elites resented paying municipal property taxes and feared that immigrant- and working-class-controlled city councils would engage in imprudent financial ventures (Elkind, 1997).

And neither provinces in Canada nor states in the United States were at all interested in giving up their constitutional power to create and abolish cities and define municipal powers. The combination of a binary constitutional arrangement that included only federal and provincial/state actors and the elite fears about municipal politicians just mentioned resulted in insurmountable obstacles being placed in the path of what in England came to be called "gas- and water-socialism" (even when political will in that direction came to the fore, which was very seldom).

At the level of formal law (including here both statute law and judicial decisions), the period from the 1870s to the 1920s was one of growing restrictions on municipal jurisdiction (Frug, 1999; Hartog, 1989; Levi & Valverde, 2006). Legislatures, courts, and authors such as municipal law expert John Dillon all worked to clip the wings of local politicians and city employees, jurisdictionally and financially. What came to be known as Dillon's doctrine (which remained good law for over a century and has been invoked by Canadian courts as well) had two key points. First, that municipalities could exercise only powers that were specifically named in appropriate state statutes ("prescribed powers"); second, that the capacious concept of "the police power of the state" could not be used – as it had been in the past – as a blanket justification and rationale by municipal governments that needed to address a new issue, govern an unexpected problem, or raise new funds (see Dillon, 1881; Frug, 1999; Levi & Valverde, 2006).

Dillon's campaign to hamstring municipalities was to a large extent successful at the level of formal law, and a similar process went on in Canada as well. However, what legal historians of the municipal have not appreciated is that local actors, both governmental and not, found it relatively easy to circumvent the court- and legislature-imposed restrictions simply by setting the legal form of the municipal corporation to one side while developing a host of "special purpose," quasi-public authorities that wielded the very powers that had been denied to elected municipal councils and then some.

In the story of how the relative powerlessness of municipal corporations came to dialectically generate an interest in and a demand for special-purpose entities that would not be subject to the same limits, and that would not be very visible to courts, legislatures, the local press, and the general public, the Port Authority of New York (later of New York and New Jersey), headed by famed local governance czar Robert Moses, is an important protagonist. In 1942, when public finance methods drew more attention than usual because of the need for war bonds, a New York municipal lawyer told a Bar Association audience that two revenue bonds issued by this body in 1926, one for $14 million and one for $20 million, constituted a "landmark" in local finance. The success of these bond issues established (in keeping with the 1898 Supreme Court *Walla Walla* decision) that such financing measures created a "self-contained debt" owed only by the authority in question – not by the city or by the state (Edelstein, 1942–3).

Municipal and state governments sensitive to taxpayer fears about public debt thus were able to quietly foster infrastructure development, without overturning state and judicial limits on city councils' power, through arm's length bodies whose debts were and are legally considered to be "off the books." Perhaps as important, such entities were and continue today to be run by appointed boards whose decisions are not scrutinized by either taxpayers or the local press in the way that municipal council decisions are. US law dictates that any authority with taxing power should be elected, but the line between a tax and a fee for service (e.g., a charge for running water) is not, in practice, easy to draw. It is easy for public authorities to define their revenue as a toll, a fee, or a price for a service or a commodity, rather than as a tax, even when there is no realistic consumer choice because the goods and services in question are provided throughout the territory of the municipality on a monopoly basis. And if the revenue flowing into the coffers of the authority in question is defined as a fee, toll, or price, then even in the no-taxation-without-representation United States, citizens will not expect to have a direct say in the governance of the public services in question.

Hybrid Governance Networks before Neoliberalism

Historically, then, moves that impose restrictions on a municipal government and decree legal and political standardization (for the municipality) have generated a largely silent response that greatly increases the number of ad hoc, non-standardized, often sui generis bodies wielding considerable power. And while no thorough study of public authorities and special-purpose bodies across issues and jurisdictions exists, so that generalizations are risky, it appears that this dialectic has continued into the present day. It is this dialectic, a large-scale historical process, that underlies and explains the creativity that has given us the countless entities that today combine the state's coercive power to tax and/or to expropriate with private capital's financing flexibility and lack of transparency and democracy.[1]

When current P3s and other more or less sui generis arrangements devised to carry out local projects of public significance are looked at in a larger context, specifically the historically informed context only briefly sketched above, it becomes possible to get beyond the largely ideological disputes about whether public powers and/or public money is better than private powers and private financing.

Special-purpose public or quasi-public authorities have remained largely unstudied, or perhaps more accurately, they have been studied only by rather narrow policy specialists (e.g., the authors included in Perry, 1995). A key reason for this unfortunate neglect is that, in North America at least, they are not products of what I would call top-down legal design. In every instance with which I am familiar, special-purpose public or hybrid authorities have emerged as site-specific and largely ad hoc answers to particular governance problems. As mentioned at the outset, in the United States, during the New Deal years especially, state-level and/or federal-level leaders wanted to meet local infrastructure needs but mistrusted city politicians, and therefore they sought alternative ways to deliver local infrastructure funding (Elkind, 1997; Foster, 1997: 18–20; Perry, 1995). Interestingly, a similar process appears to be occurring now in China, where the same central government that banned municipal borrowing because of corruption concerns has had to create "alternative investment vehicles" so that urban infrastructure work can continue.[2]

While in some historical cases such special-purpose entities have been deliberately created by higher levels of government (as was the case in Thatcher's Britain [Deakin & Edwards, 1993; Lawless, 1988]), in many other cases "higher" levels of government were wholly uninvolved. Urban parks systems are a case in point. Local elites – whether from mistrust of elected local politicians, a desire for personal control over a specific public activity, and/or a desire to establish entities that were worthy recipients of philanthropy – pushed for and obtained ad hoc local governance bodies run by gentlemen and ladies like themselves (through boards), often supplemented by experts such as architects or engineers. Jon Teaford, the leading historian of local government in the United States, points out that suspicion of immigrant- and/or working-class city politicians combined with an interest in aesthetic philanthropy as the Illinois legislature created not one but three different parks boards for Chicago in the late nineteenth century.

These boards, which garnered considerable cultural capital for elites of the "robber baron" era, were allowed not only to borrow money that did not count towards the municipal debt but even to levy taxes (Teaford, 1984: 68). Such entities have relied on the old, pre-democratic legal mechanism of the appointed or self-perpetuating board of directors or board of trustees.[3] Such arm's-length but financially comfortable "public" library boards and parks boards persisted even after the philanthropic influence of the Andrew Carnegie variety waned and financial responsibility for such endeavours was assumed by cities. New York's Central Park, for example, is currently run not by the city parks department but rather by an independent non-profit group, the Central Park Conservancy, which has its own governing board. This entity describes itself as being under contract to the City of New York (see Central Park Conservancy, 2016).

Boards charged with establishing, funding, and governing specific parks or libraries and even whole municipal parks systems or library systems; harbour (and later, airport) authorities; water districts, which may or may not be coterminous with cities but are (in the United States, at least) often legally and financially independent of the legal city; conservation districts (and, in the province of Ontario, conservation authorities); entities such as the New Jersey Turnpike or the Triborough Bridge in New York, which are not for-profit corporations but are not subordinated to any elected local government – even without proceeding to provide a more exhaustive list, it is clear that there is little or nothing that such entities have in common either in their mission or in the political motives of their creators. However, one common denominator (from the governance point of view) is the circumvention, intended or unintended, of the democratic processes and accountability mechanisms characteristic of modern democratic local governments. Even when city councillors sit on these boards, the work they do on them is not visible to the constituents in the way that regular city council business is, and in many instances city councillors are outnumbered by appointed members whose qualifications are rarely revealed to local citizens.

Now, I am not claiming that the philanthropists who set up parks and library boards or the bureaucrats who set up port and highway authorities were deliberately trying to evade democracy. While this may have been the case in some instances, it is likely that in the vast majority of cases new institutional forms came into being for pragmatic reasons: it just "made sense" to set up a special-purpose authority, and once set up, there were certain governing habits that, again, "made sense." Thus, it might be more cautious to speak not about countermoves against local democracy but rather about countercurrents.

Special-purpose authorities live in the political shadows, emerging into public view only when there are spending scandals or major jurisdictional disputes (e.g., in Toronto, the longstanding war between the federal and the local government in regard to the Port Authority's plans for the city's Island Airport). Sporadic local media coverage of occasional interjurisdictional or purely local bunfights does not help to shed much light on the general workings of the political-legal-informal governing assemblages that underpin the formally existing entities that in turn build and operate many key local infrastructures.

In the United States the actions of school boards and police boards/ commissions have certainly been studied, though usually with an exclusive focus on racial effects, but the governance of other types of local service has drawn very little attention (for exceptions see Foster, 1997; Perry, 1995). Information on special-purpose authorities is, in fact, plentiful, but it is very scattered and usually not analysed in governance terms. For example, Susan Fainstein's influential study (2001) of urban development politics in London and New York contains much information about English quangos and US urban development corporations, but the governance entities discussed are not placed in any larger legal context and are discussed as if the legal forms used were invented for the first time by those studied. The absence of historical and comparative perspective in this deservedly famous book is a symptom of the problem underlined here: the lack of comparative overviews of different ways of building and running local public services across jurisdictions.

An exception is the growing literature on one small section of the broad governance field under study here – that section composed of partnerships between government entities and for-profit corporations (P3s). Administrative law scholars as well as planners have certainly sounded alarm bells in regard to P3s, both in relation to value for money and in relation to governance and accountability. However, this does not mean that, if P3s were abolished, the public interest would be significantly better served – even though that is the assumption that underlies much of the relevant critical literature on P3s from both administrative law and urban planning (see Custos & Reitz, 2010; Freeman, 2011; Hodge, 2004, 2006; Sagalyn, 2007; Siemiatycki, 2010). The warnings about the effects of using contracts rather than public law, by-laws, and regulations are well taken. But especially given the long and sad history of exclusionary zoning and class-biased planning in North America, it is problematic to assume that any use of public law and public legal forms is by definition better than any use of private law or quasi-legal techniques and forms. Thus, the assumption that governing urban infrastructure through contracts and contract law rather than through public legal tools is necessarily regressive and neoliberal may fit the facts in many instances, but not necessarily all. In other words, public law is not always more democratic in its effects than private law, contrary to the core assumptions of administrative law scholars.

One instance where a contract proved a better way to serve the public interest than a public legal tool is found in a current Toronto district-sized development, the West Donlands. Those managing the "deal" with developers, representing the special-purpose authority Waterfront Toronto, inventively decided to set out the maximum height of future buildings in the contract. Given that in Ontario there is a long history of municipally legislated height and density limits being overturned by the higher-level, appointed tribunal, the Ontario Municipal Board, the use of a contract instead of a zoning law was nothing short of brilliant; contracts, obviously, are not subject to the usual zoning appeal process by which developers go over the city's head and obtain higher densities from the provincial Ontario Municipal Board (Bunce, 2011). The example demonstrates that administrative lawyers' and planners' professional preference for public law tools may sometimes blind them to creative uses of private law.

Similarly, while a proliferation of one-off deals and authorities with unique powers is bound to pose risks for accountability and transparency, it does not mean that the standardization of local legal forms (the general project most famously embodied in nineteenth-century general municipal incorporation statutes) is inherently and consistently democratic. After a good century or so of municipal democracy, it has become clear that the formal-law move that replaced the unique privilege-conferring city or borough charters with standard municipal incorporation (Frug, 1999; Isin 1992, Williams, 1985), even when combined with the related move that replaced the property franchise by universal adult local suffrage, has not in practice sufficed to generate either good government or real democracy – as Martin Loughlin has shown in his historically informed analysis of English courts' interpretation of municipal "fiduciary duty" (1996: chaps 4 and 6). The political dominance of the ratepayer identity and homeowner discourse in local politics means that democracy on paper does not necessarily generate real democracy, especially for tenants and the marginally housed.

Because the municipal corporation has not been critically studied in a systematic fashion either by legal historians or by urban studies scholars,[4] a comparative legal history of municipal incorporation in the common law world is not feasible at this time. It is thus even more challenging to tackle the even more difficult task of mapping the complex and heterogeneous urban governance networks within which the municipal corporation has been, in fact, located. But it is nevertheless possible to use the available fragmentary evidence to begin sketching a legal genealogy of urban governance that does not take the legal form of the standardized modern municipal corporation for granted.[5]

Legal historians have established that in the mid-nineteenth century great efforts were made, especially in the United States but also in Canada and England, to impose a standardized municipal corporation *format* on urban governance (Isin, 1992; Loughlin, 1996: chap. 1; Webb & Webb, 1908; Williams, 1985). There was also a great and partially successful effort to separate "public" from "private" interests and public from private corporations (Hartog, 1989; Horwitz, 1982; Kaufman, 2008) and, relatedly, to deprive municipal corporations of financial resources and powers. The combination of these two legal-political processes had the effect, intended or not, of forcing municipalities to become increasingly dependent on rates (local property taxes) and on discretionary government grants. The same process of legislative and judicial pushing and shoving also made the municipal corporation more predictable as a legal form (in contrast to the highly specific charters held by many older cities). This was certainly an intended effect, as one sees especially in England, where there was a virtual moral panic about "rotten boroughs" and privileged municipal corporations in the wake of the 1832 Reform Bill. But what is not generally known is that the Herculean effort to render urban legality more predictable and more legible from above had much more success at the level of doctrine than it did in practice. Only one-seventh of England ended up being covered by the Municipal Corporations Act of 1835, and even in spaces in which a standard-issue municipal corporation existed, a vast array of transportation-, water-, and conservation-related authorities still existed. The same situation existed in the United States (as we saw earlier in the example of the proliferation of special-purpose bodies during the New Deal).

In rewriting the history of urban governance "as it really was" – rather than as courts and legislatures decreed it should be – it is dangerous to assume that the proper vehicle for the people's welfare at the local level is always and everywhere public law, as mentioned above in relation to the Waterfront Toronto contracts. But by the same token it is necessary to beware of the Gierkean temptation: the belief that if the legal form of the municipal corporation has become regressive (as Gerald Frug has famously showed for the United States), the solution lies in reviving a pre-legal cohesive local community, an organic association (Frug, 1999; Gierke, 1951).

In general, a great deal of ink has been spent waving a flag for one or another of the entities found among local governance assemblages – the public municipal corporation; private enterprise; the organic local community. By contrast, too little ink has been spent documenting varieties of local governance assemblages in their rich pluralism.

The legal history of local governance is not a well-developed field, generally. But for the United States, at least, there is plenty of evidence that "the city," in the legal sense of city council, has rarely engaged in "seeing like a state" planning and has often led from behind (Elkind, 1997; Frieden & Sagalyn, 1989). American downtown redevelopment schemes, from the Great Depression onward, have consistently been led by the private sector (including, again, civic leaders who were not motivated by profit, but were not part of government either [Isenberg, 2004]) and governed by entities, the predecessors of today's urban development corporations, that were, in a sense, P3s.

That "partnership" is often a misleading term when applied to infrastructure projects has been repeatedly noted in the relevant policy literature, since many have pointed out that in law a partnership is a great deal more than a contract or even a bundle of contracts (Custos & Reitz, 2010; Hodge, 2004, 2006; Siemiatycki, 2010). However, while noting that "partnership" is inaccurate in that its long-term financial interdependence is not necessary or even common in infrastructure "partnerships" (which are often leases, concessions, or contracts rather than partnerships), the underlying binaries – public law vs. private law, public interest vs. private interest – are generally taken for granted in the literature, by the left as well as by the neoliberals. The black-boxing of these fundamental liberal political binaries has prevented us from closely studying the actual workings of the relevant legal forms – from the municipal corporation to the construction contract.[6]

One way to theorize the kinds of entity that are the focus of this chapter in a non-normative or post-normative manner would be to look at the interaction of their *jurisdiction(s)* on the one hand and their *scale* of operation on the other (cf. Valverde, 2009). Their geographic scale is often quite limited. The Triborough Bridge in New York, for example, is not nearly as big, physically, as the hugely complex governance assemblage that built it. But, by contrast, their jurisdictional reach, or perhaps more accurately, their *jurisdictional depth*, can be immense. In many cases these special-purpose entities are allowed to set their own planning rules as if they were micro-municipalities, even though they also own the land and/or the buildings (a conflict of interest that planners rarely remark upon). They can also hold and sell land independently of the municipal government, and (often) they can borrow money that does not count as part of the municipal or state debt, which in the United States and Canada, at least, is often a key factor in their creation and in their persistent popularity. They also often enjoy the power of expropriation. Thus, if one looks only at the physical space covered, they look relatively small, but the way in which public powers and private sector *capacities* and *legal powers* are piled on top of one another is anything but "micro."[7]

Calling these entities neoliberal innovations is misleading because these bodies are also the direct and legitimate offspring of the collective economic actors, simultaneously wielding public powers and making private profits, that emerged in the Middle Ages. These "associations" (to use Gierke's term) include the medieval abbeys (one of which, Cluny, can claim the title of first transnational corporation [Berman, 1983]); the English manors; Oxford and Cambridge universities and their colleges; the corporation of the City of London; many charitable trusts and foundations; and on the more commercial end, the Hudson's Bay Company (which not untypically was also a P3, being granted policing and criminal justice jurisdictions as well as a monopoly on the fur trade in its 1670 charter).

The pluralism of local governance is thus not limited to the geographical and jurisdictional overlap of the many authorities that converge on particular urban spaces or projects. That type of pluralism is important: what inhabitants call "the city" is in practice a collection of disparate authorities, including school districts, conservation authorities, parks boards, public housing authorities, police force structures, and so on. These sub-municipal or "inter-municipal" entities have specific interests as well as jurisdictions and in many cases have boundaries that do not match those of the municipality.

In addition to that pluralism there is an internal division within many of the local authorities just listed, one signalled by Hartog's purposively paradoxical phrase "public property and private power" (1989). The mix of jurisdiction and property that Hartog identified as constituting the old municipal corporation did not, in fact, disappear in the nineteenth century, despite the work of modernizers who sought to render local government more predictable and legible and to ensure that public corporations did not become serious competitors to private enterprises.

It is high time to question the validity of the well-established claim made by Hartog and other American legal historians that property and jurisdiction were clearly separated in the mid- and late nineteenth century, and that property was successfully taken away from municipal corporations while local government power was in turn confined to the municipal corporation. The reality is far more complex. Municipal corporations today still hold property and extract revenue from it; the reality that selling off municipally owned buildings has become popular in neoliberal times speaks to the fact that cities own considerable amounts of real estate in our own time. By the same token, many entities that are not part of government and are run neither by elected officials nor by government contractors continually deliver all manner of public services and goods. The governance of local infrastructure is not properly understood if we imagine that such work was in the past purely governmental and is now being either privatized or subject to public-private partnerships. Hybrid and intermediate entities have long been crucial actors in the delivery of infrastructure and local public services. Neoliberal practices and habits of governance are important, of course, as countless critical urban scholars have shown (e.g., Kipfer & Keil, 2002); but ascribing a great deal of agency to neoliberalism in general when confronted with something called a "partnership" is both empirically and theoretically dangerous.

However sketchy, this discussion of the genealogy of public-private partnerships and the long and largely unknown history of public authorities, special district governments, and other hybrid local governance entities shows that much is to be gained if urban studies and legal scholars turn away from the usual normative questions about where the famous "line" between "the public and the private" ought to be drawn and turn instead to historically informed inquiries from which analytical and theoretical lessons can be drawn. Thus, this chapter takes issue with choosing "neoliberalism" as the object of analysis – since I believe that focusing on concrete governance networks is much more fruitful than prejudicing the inquiry by presupposing that what one is studying is "neoliberalism."

Therefore, in keeping with earlier work (not at the local scale) on what I then called "the mixed social economy" (Valverde, 1996), I would suggest that it is time for scholarship on urban governance to proceed as follows:

1. Instead of asking "is there privatization?" and "is it bad?" it would be more fruitful to ask the following questions.
2. What are the resources deployed in a specific project, and where do they come from?
3. What jurisdictions and other legal powers are at work in specific governing assemblages?
4. What is the de facto authority being exercised by which groups or individuals?
5. What personnel is regarded as appropriate?
6. What are the techniques used to build the governing assemblage in question, including legal forms such as the contract? Were these techniques borrowed and repurposed, and, if so, what kinds of governing effects trail in their wake?

These questions hold more promise than the standard binary questions counterposing "the public" in general to "the private." The above sketch of the history of urban governance assemblages has shown that, when asking those questions, it is wise to set aside one's assumptions about whether public law is more or less virtuous than private law and about whether the public interest is always better served by public servants and public law tools rather than by hybrid and unique governance assemblages.

NOTES

1 The term "creativity" is not meant to have a positive normative value, since it could be argued that the creativity displayed by those who design and create special-purpose authorities may be due to a desire to avoid following the rules that over the years have been imposed on municipal councils, including the unionization of the municipal workforce and rules promoting transparency and accountability.
2 Personal communication from Professor Mariana Prado (University of Toronto Law School).
3 The use of the term "board of trustees" does not always indicate the existence of a legal trust; similarly, the use of the term "board of directors"

does not always indicate the existence of a non-profit or public corporation. That is, people who serve on boards of trustees may or may not be bound by the law of fiduciary duty. Other terms may also be used; for example, the University of Toronto, formally an Ontario public, non-profit corporation, is run by a "governing council" whose members are called "governors" or "members of governing council," but other universities in the same province have "boards of trustees."

4 Gerald Frug, Richard T. Ford, and Yishai Blank, among other scholars, have contributed to such a project, but I am unaware of any synthetic work that does for the municipal corporation what so many critical political theorists have done for "the state." Engin Isin's (1992) highly original interdisciplinary critique of the non-democratic effects of the standardization of the municipal corporation form is a rare exception, but the book, published by an obscure Montreal publisher, has not been taken up and developed by later scholars.

5 The modern standard-issue municipal corporation looked like a great idea in the England of the 1830s, in contrast to the much publicized evils of close corporations, "pocket boroughs," and so on. But while the modernizers' denunciation of local privileges and irrationalities served a useful democratizing as well as rationalizing purpose in its day, scholars who now cut and paste from such frankly biased sources as the Webbs' account of the evils of local heterogeneity in pre-1835 English local government are unwittingly hampering the study of actually existing networks of local governance. (Whether the Webbs' horrified attitude towards local governance heterogeneity was one of the reasons why they became staunch supporters of the Soviet Union later in life is a question worth pondering).

6 The term "legal forms" is used to try to avoid the blinkers that are inherent in legal scholars' tendency to focus on either standard-issue institutions or established legal tools. "Legal forms" is a purposively capacious term that encompasses legal tools that are used across a wide variety of institutions, such as contracts, zoning by-laws, or public regulations issued pursuant to a statute, but it also covers constitutionally visible generic institutions such as the municipal corporation and the for-profit limited liability company. But the term can also cover what one might call infraconstitutional entities, such as the boards of directors that run non-profit companies, both governmental and community based, as well as the even more infraconstitutional and indeed infrajuridical processes that are relatively established but are merely habitual or informal. (The word "infraconstitutional" is adapted from Fleur Johns's [2013] term "infralegal.")

7 In analysing, in a non-legalistic way, the workings of local governance
   assemblages I am influenced by Saskia Sassen's (2006) terminology;
   rejecting the usual division of governance into global, national, and
   local spheres, she shows that local capacities can often be used for global
   agendas and, and that powers found at one "level" can easily be used to
   further interests that exist at another level. My analysis of the complex
   relations between jurisdiction and scale has affinities with her call for "new
   jurisdictional geographies" (388).

REFERENCES

Berman, H. (1983). *Law and revolution: The formation of the western legal tradition.*
   Cambridge, MA: Harvard University Press.
Bunce, S. (2011). Public-private sector alliances in sustainable waterfront
   revitalization. In G. Desfor & J. Laidley (Eds), *Reshaping Toronto's waterfront*
   (287–304). Toronto: University of Toronto Press.
Central Park Conservancy (2016). About us. http://www.centralparknyc.org/
   about/about-cpc/. Accessed 7 March 2016.
Custos, D., & Reitz, J. (2010). Public-private partnerships. *American Journal of
   Comparative Law, 58*(0), 555–84. http://dx.doi.org/10.5131/ajcl.2009.0037
Deakin, N., & Edwards, J. (1993). *The enterprise culture and the inner city.*
   London: Routledge.
Desfor, G., & Laidley, J. (Eds) (2011). *Reshaping Toronto's waterfront.* Toronto:
   University of Toronto Press.
Dillon, J. (1881). *Commentaries on the Law of Municipal Corporations.* 2 vols.
   Boston: Little, Brown.
Edelstein M (1942–3). *The authority plan: Tool of modern government. Cornell Law
   Quarterly, 28*, 177–98.
Elkind, S. (1997). Building a better jungle: Anti-urban sentiment, public works,
   and political reform in American cities, 1880–1930. *Journal of Urban History,
   24*(1), 53–78. http://dx.doi.org/10.1177/009614429702400103
Fainstein, S. (2001). *The city builders: Property, politics and planning in London
   and New York.* Lawrence: University of Kansas Press.
Foster, K. (1997). *The political economy of special-purpose government.*
   Washington, DC: Georgetown University Press.
Freeman, J. (2011). The contracting state. *Florida State University Law Review,
   28*, 155–200.
Frieden, B., & Sagalyn, L. (1989). *Downtown Inc.: How America rebuilds cities.*
   Cambridge, MA: MIT Press.

Frug, G. (1999). *City making: Building cities without building walls*. Princeton: Princeton University Press.

Gierke, O. von (1951). *Political theories of the middle age* (1st ed. 1900). Cambridge: Cambridge University Press.

Hartog, H. (1989). *Public property and private power: The corporation of the city of New York*. Ithaca: Cornell University Press.

Hodge, G. (2006). Public-private partnerships and legitimacy. *University of New South Wales Law Journal, 29*(3), 318–27.

Hodge, G.A. (2004). The risky business of public-private partnerships. *Australian Journal of Public Administration, 63*(4), 37–49. http://dx.doi. org/10.1111/j.1467-8500.2004.00400.x

Horwitz, M. (1982). The history of the public-private distinction. *University of Pennsylvania Law Review, 130*(6), 1423–8. http://dx.doi.org/10.2307/3311976

Isenberg, A. (2004). *Downtown America: A history of the place and the people who made it*. Chicago: University of Chicago Press. http://dx.doi.org/10.7208/chicago/9780226385099.001.0001

Isin, E. (1992). *Cities without citizens: The modernity of the city as a corporation*. Montreal: Black Rose Books.

Johns, F. (2013). *Non-legality in international law: Unruly law*. Cambridge: Cambridge University Press.

Kaufman, J. (2008). Corporate law and the sovereignty of states. *American Sociological Review, 73*(3), 402–25. http://dx.doi.org/10.1177/000312240807300303

Kipfer, S., & Keil, R. (2002). Toronto Inc.? Planning the competitive city in the new Toronto. *Antipode, 34*(2), 227–64. http://dx.doi.org/10.1111/1467-8330.00237

Larner, W., & Butler, M. (2007). The places, people and politics of partnership. In H. Leitner, J. Peck, & E. Sheppard (Eds), *Contesting neoliberalism: Urban frontiers* (71–89). New York: Guilford Press.

Lawless, P. (1988). Urban development corporations and their alternatives. *Cities (London, England), 5*(3), 277–89. http://dx.doi.org/10.1016/0264-2751(88)90046-7

Levi, R., & Valverde, M. (2006). Freedom of the city? Canadian cities and the quest for governmental status. *Osgoode Hall Law Journal, 34*(3), 409–60.

Loughlin, M. (1996). *Legality and locality: The role of law in central-local government relations*. Oxford: Clarendon Press. http://dx.doi.org/10.1093/acprof:oso/9780198260158.001.0001

Novak, W. (2001). The American law of association. *Studies in American Political Development, 15*(2), 163–88. http://dx.doi.org/10.1017/S0898588X01000037

Perry, D. (1995). Building the city through the back door: The politics of debt, law, and public infrastructure. In D. Perry (Ed.), *Building the public city: The politics, governance and finance of public infrastructure* (203–14). Thousand Oaks, CA: Sage.

Rose, W.A. (1953). Developments in revenue bond financing. *University of Florida Law Review, 6*, 385–90.

Sagalyn, L. (2007). Public-private development: Lessons from history, research, practice. *Journal of the American Planning Association, 73*(1), 7–22. http://dx.doi.org/10.1080/01944360708976133

Sassen, S. (2006). *Territory, authority, rights: From medieval to global assemblages.* Princeton: Princeton University Press.

Siemiatycki, M. (2010). Delivering transportation infrastructure through public-private partnerships: planning concerns. *Journal of the American Planning Association, 76*(1), 43–58. http://dx.doi.org/10.1080/01944360903329295

Teaford, J. (1984). *The unheralded triumph: Municipal government in the US, 1870–1900.* Baltimore: Johns Hopkins University Press.

Valverde, M. (1996). The mixed social economy as a Canadian tradition. *Studies in Political Economy, 47*, 33–49.

Valverde, M. (2009). Jurisdiction and scale: Using technicalities as resources for theory. *Social & Legal Studies, 18*(2), 139–57. http://dx.doi.org/10.1177/0964663909103622

Webb, S., & Webb, B. (1908). *The manor and the borough.* 2 vols. London: Frank Cass.

Williams, J.C. (1985). The invention of the municipal corporation: A case study in legal change. *American University Law Review, 34*, 368–431.

PART 4

Neoliberal Welfare and Philanthropy

10 Governing through Failure: Neoliberalism, Philanthropy, and Education Reform in Seattle

KATHARYNE MITCHELL AND CHRIS LIZOTTE

Led by large-scale corporate foundations such as the Bill and Melinda Gates Foundation, Walton Family Foundation, and Broad Foundation, philanthropic interventions in education have become a dominant aspect of the landscape of public school reform in the United States over the past fifteen years (Kovacs, 2011; Ravitch, 2010, 2013; Reckhow, 2012; Saltman, 2010; Scott, 2009). While the ideological motivations of donors vary, the desired results have tended to converge towards the creation of "quasi-market," choice-based mechanisms within the public education system. Typical funding support is directed towards non-professional teaching corps programs,[1] pedagogic experimentation in school districts, education reform advocacy organizations (ERAOs), non-profit and for-profit charter management organizations (CMOs),[2] and, in some cases, direct and in-kind donations for political campaigns aimed at deregulating the public schooling "market" (see Cohen & Lizotte, 2015).

Philanthropic organizations in the United States have long been interested in establishing or bolstering educational institutions such as libraries and universities as a means of developing human capital and improving society (Lagemann, 1989; Zunz, 2011). Today's "new" philanthropy, known variously as venture philanthropy, or philanthrocapitalism, similarly embraces the liberal rationalities of human capital development, but also employs incentives that nudge recipients towards neoliberal practices of entrepreneurialism, calculation, and prudentialism (Mitchell & Lizotte, 2014). Many contemporary foundations insist on competitive grant application processes; standardized systems of assessment and social evaluation; targeted, short-term projects that can be scaled up if successful; and some form of return on investment (Bishop & Green, 2008; McGoey, 2012). In the realm of primary and secondary education these parameters often favour actors and institutions that propose or undertake projects emphasizing more individualized parent, teacher, and/or administrative responsibility for student outcomes; quantitative metrics assessing teacher and student performance; and/or technologies of expertise (e.g., standardized tests and consultants' reports) in proposed reforms (Kovacs, 2011; Saltman, 2010).

The participation of corporate philanthropy in political struggles around public education raises many questions, but in this chapter we focus on just one: how actors on the ground play a critical role in the development and transformation of neoliberal practices and rationalities such as these in the educational arena. As Peck (2010) and others have argued, one of the most remarkable characteristics of neoliberalism is its general variability and adaptability (see also Brenner, Peck, & Theodore, 2010; Leitner, Peck, & Sheppard, 2007; Ong & Collier, 2004). These theoretical insights often draw heavily on the empirical research of geographers, sociologists, and anthropologists engaged in ethnographic on-the-ground work around the globe (e.g., Atia, 2013; Brady, 2011; Li, 2007; Lippert, 2005; Roy, 2012a). "Actually existing neoliberalism" is shown by these studies to consist of often disconnected and contradictory practices, always indicating both historical forms of path dependence and geographical adaptation to local context. This may be particularly true vis-à-vis the assemblage of governmental ensembles, where non-liberal and liberal rationalities converge and take shape in specific ways depending on the socio-spatial milieu (see Foucault, 2008; Dorow, Larner, Li, Lippert, and Valverde in this volume).

What the geographic perspective reveals particularly well is that neoliberal practices are differently efficacious or "successful" vis-à-vis their multiple applications across space. Certain neoliberal projects take root and are sustained more easily than others, owing to local political, cultural, and social factors, while some fail to be adopted or fade away after a short time. It is the wealth of data generated by these variable successes and failures that leads to what Peck, Theodore, & Brenner (2010) refer to as neoliberalism's capacity to learn and adapt. Failures as well as successes are transformed into knowledge circulated among policy makers and practitioners as "best practices" to be emulated or "pitfalls" to be avoided (McCann, 2011).

We demonstrate this aspect of the larger phenomenon of learning and transformation in neoliberal forms of governance by investigating one example of philanthrocapitalist intervention in education reform. To date most studies of philanthropic involvement in education have portrayed philanthropy as a monolithic, omnipotent force capable of imposing its desired changes upon schools, communities or entire districts (Saltman, 2010). The reforms that are studied are seen to take place through funding contingent on draconian changes in educational policy such as mayoral takeovers of districts or forced conversions of neighbourhoods to charter schools.

While the impact of many educational changes imposed from above is unquestionable and has been the subject of several excellent studies (e.g., Lipman, 2011; Ravitch, 2010; Reckhow, 2012; Scott, 2009), not as much research has focused on how philanthropic organizations actually interact with community partners to achieve or adapt their desired outcomes. This chapter draws on a ground-level study, conducted between November 2010 and October 2011, of a major philanthropic organization and its repeated attempts to influence education policy in Seattle, USA. The emphasis here is on the inertia, resistance, and blockages to change and the ensuing transformation in the foundation's granting policies that resulted from the unexpected outcomes of the process of local political engagement. In doing so, the study shows that philanthropic involvement in education reform is not a straightforward process of incentivizing and hence inculcating ideas of choice, entrepreneurialism, calculation, assessment, prudentialism, or other forms of neoliberal rationality in administrators and institutions. Rather, ground-level actors have an important, if largely unexplored, role in the uneven development of neoliberalism across space and time.

More specifically, we show how the Bill and Melinda Gates Foundation (hereafter the Gates Foundation) attempted to undertake a partnership with Seattle Public Schools to create a series of academically themed "small learning academies" in the city's high schools. The ultimate failure of the project led the Gates Foundation to radically change its approach to urban school reform, including shifting its partnership with local public school districts to more reform-friendly intermediary organizations operating at the national level. In addition to demonstrating the capacity of neoliberal policy-promoting institutions to adapt and learn, the case study also contributes to this volume on neoliberal governmentality by showing how ethnographic methods can illuminate how practices associated with neoliberal governance are mobilized locally and how they evolve at different levels. In addition, this work adds nuance to what has until recently been a fairly monolithic view of philanthropic involvement in school reform by demonstrating that, despite overwhelming political and financial capital, corporate philanthropy cannot simply force change, but rather must interact with local milieus.

This study is based on archival information dating to 2011 and interview data collected from Seattle-area education reform stakeholders and school district officials. Archival data were collected from various media, government, and private sources. Interview participants were identified using snowball sampling on the basis of recommendations of a select number of initial contacts. With one exception noted below, interviews between the researchers and informants were transcribed from original audio recordings and informants were given pseudonyms in this text, where appropriate, to maintain their anonymity.

We close this introduction by noting that, as a single case study of a specific location, we do not mean to suggest that what happened in Seattle is emblematic of either school reform or neoliberal governmentalities more generally. Rather, we want to emphasize the many disarticulations in the process of governance that a particularly influential and powerful philanthropic organization encountered locally and how it adapted and evolved in the face of this "failure." Failure, as we employ it here, is a frequently productive force, one that has the capacity to propel new programs forward as well as to enable the adaptation of existing ones (cf. Hunt & Wickham, 1994). We agree with early governmentality work that it is often around "difficulties and failures that *programs of government* have been elaborated" and through which organizations "seek to configure specific locales and relations in ways thought desirable" (Rose & Miller, 1992: 181; emphasis in the original). Where we disagree is in the apparent seamlessness of these elaborations and articulations, especially with respect to the translation of the programs into the consciousness and practices of individuals and groups.

Thus, while we recognize that the literature on neoliberal governance has a great deal to say about the subjects of governmentality and their recasting as self-governing citizens of a neoliberal world, we believe there is inadequate discussion about how this change is actually accomplished (or not) in practice (cf. Brady, 2014; Collier, 2012; Hilgers, 2010; Wacquant, 2012). Therefore, we want to add to the growing body of empirical work signifying how Foucault's "technologies of the self" (2008) and other principles of personal and institutional entrepreneurship, choice-making, calculation, and self-care are inevitably and importantly formed and reformed *in place* as they intersect with the mundane realities of local political and cultural circumstance.

Geographies of Neoliberal Governmentality: The Context of Context

The compelling evidence that variation and unevenness characterize reforms associated with neoliberalism immediately puts pressure on monolithic time-space meanings of the term. Focusing a geographic lens on the processes characteristic of neoliberal governmentality creates investigations that are contextually nuanced and locally specific. But at the same time, it enables a view of the structuring and restructuring processes that generate patterns in the urban and political landscape over time.

Contextual socio-spatial approaches to neoliberalization are thus well positioned to observe the *both/and* nature of these framing processes: they can be structurally oriented *and* contextually nuanced, broadly patterned *and* locally specific, and successfully market disciplining *and* effectively contested. This type of analytical lens enables a view of neoliberalization as something that is constantly evolving and adapting, but not in an ad hoc, random, or purely contingent way. Rather, it can be seen as a process that develops through regulatory experimentation, the circulation and hybrid formations of ideas and rationalities (see Howard in this volume), and relational interactions – involving both human and non-human actors. Brenner et al. (2010: 190) wrote of this patterned yet non-linear sequence of interactions: "It follows that the (uneven) neoliberalization of regulatory arrangements across places, territories and scales qualitatively transforms the institutional landscapes and interspatial policy relays within which subsequent neoliberalization projects are mobilized."

This approach emphasizes the importance of cumulative impacts over time, the historical and geographical sequence of events that transform institutional landscapes and set into motion succeeding rounds of transformation and restructuring. These ideas build on the work of geographers Neil Smith (2008) and David Harvey (2006), who have emphasized the patterning of regulatory landscapes reflective of the production of uneven development through time. The intensification of uneven development in spatial landscapes interacts with the formulation of new institutional or political regimes of governance; these transformations in turn lead to successive rounds of restructuring and market disciplining and the formation of new subjectivities and interactions. Each round of development and restructuring thus makes sense and can be understood only in relation to previous rounds. In this context it is important to note the capacity (and necessity) of neoliberalism to adapt and change and to highlight how these changes are always in relation to previous struggles and forms of consciousness and action that have been productive of institutions, places, and people through time.

Although these theorists have foregrounded structural and institutional commonalities such as economic optimization, commodification, and market logics and their role in the patterning and framing of regulatory landscapes, it is equally critical to address the role of actors and political coalitions operating both inside and outside the above processes and interactions. Geographically informed ethnographies such as those by Moore (2005), in his majestic account of territorial struggles over resettlement and dispossession in Zimbabwe, are able to highlight the key actions, beliefs, and *emotions* of those whose "suffering for territory" over time is also central to the outcome of neoliberalization and colonial capitalism *in place*. Similarly, Hart's (2002: 14) comparative ethnographic account of agrarian change in South Africa, Taiwan, and Mainland China helps manifest the convergence of socio-spatial formations (e.g., of globalization) and different axes of power such that "objects, events, places and identities ... are formed in relation to one another and to a larger whole."

Through their focus on situated practices and local context – including the vagaries of racial, postcolonial, and gender struggles as well as class politics – these scholars draw attention to the local embeddedness of regulatory restructuring. At the same time, however, they do not neglect the wider processes of uneven development, the macrospatial institutional landscapes in which neoliberalizing experiments are initiated and implemented. This is what Brenner et al. (2010: 202) call the "context of context" – the broader conditions and enframing in which local projects unfold. In investigations of the framing rationalities of neoliberal governance we want to argue in a similar manner for attention to the micro practices of situated actors, but within this broader context of market-driven reform through time. This is one piece of a more comprehensive effort at "getting Marx and Foucault into bed together" (cf. Hunt, 2004), drawing on different dimensions of their work that prove fruitful for ours (see Marston in this volume).

In the contemporary literature on neoliberal governmentalities there are increasing calls for moving beyond analyses and research conceptualized "at a distance" – for example, those relying primarily on texts and archives. A strong scholarly interest in the agenda of neoliberal governance enables (and possibly inclines) researchers to see the *concerns* of governors; the methods, practices, and domains through which actors are *able* to be governed: the conditions laid down, the incentives provided, the targets delineated, the choices proffered. But focusing on the enframing rationalities, how domains are "made thinkable and practicable by governors as ... knowable and administrable" (Rose, O'Malley, & Valverde, 2006: 86) may, simultaneously, lead scholars to miss the moments when these neoliberal rationalities fail to articulate or connect with the intended recipients. Brady (2014: 14) similarly writes of the importance of redirecting attention so as to capture these moments of "failure": "Attentiveness to the dynamics of social life within a particular group or space (or both) also supports greater attention to the processes through which political alliances are formed, as well as resistances to such alliances and new programs of governance, and the failures of various plans."

This type of attentiveness, however, demands getting down and dirty – into the messiness of human relations and of human/non-human interactions. Compliance and resistance can even be found in a single subject position, a seemingly paradoxical "*both/and*" form of articulation that cannot generally be seen in the view from above. It is critical to investigate how these types of positions and ideas are formed and mobilized and moreover how they circulate and enmesh people and places in multiple, *often contradictory* positions within market societies (cf. Roy, 2012b).

In terms of researching the domains of neoliberal governance it is quite possible to observe the setting of conditions "from above" – the provision of incentives and the nudging of behaviour – but then to miss the response, often something more complicated and inchoate. Similarly, the agents of governing "from below" are also frequently unnoticed or disregarded as they offer their own visions and assessments of expertise, risk, and care. As Lippert & Stenson (2007: 1) note, "their strategies tend to be overlooked or dismissed as integral facets of how ... governance works and – sometimes – fails to work." Yet the politics of urban or local governance is critical to understand as it often works back on the governors – whether they are trustees of improvement projects (cf. Li, 2007), philanthropists, politicians, capitalists, or some hybrid of these positions of tutelage and sponsorship.

In education, for example, neoliberal governance encourages parental and administrative choice and responsibility in the context of expert
knowledge and risk assessment. The assumption of philanthropists and
other actors interested in educational reform is that if the appropriate
expertise (and funding support) vis-à-vis schools, teachers, children,
and optimal learning methods is provided, parents and superinten-
dents will be nudged not just to make the *right* choices in these areas,
but also to become *believers* in choice and in reform more generally.
Thus, the neoliberal rationalities of education reform are such that the
technologies of expertise – standardized tests, consultants' reports, and
other metrics of expert evaluation – must be provided and encouraged
along with the appropriate financial incentives. With this expertise and
support from above the expectation is that parents and other actors
such as school boards, administrators, and teachers will then make the
appropriate prudential calculations and pursue the most rational (and
least risky) choice of action for their children, schools, and districts.

What ethnographies of education can show us is whether or not
rationalities of expertise and choice have actually connected or "taken
hold" and, if so, in what ways. Moreover, if they are unsuccessful or
only partially successful, what are the forms of refusal that have been
made, how, by whom, and, perhaps most important, what are their
material effects? In the next section we examine the logics of responsibility, choice, risk, and assessment in education that are part and par-
cel of the broader forms of neoliberal governmentality now saturating
many contemporary institutions in the United States. We then turn to
a case study of the Gates Foundation, which has been a key player in
incentivizing these newer rationalities, but which has encountered the
sticky particularities of place-based knowledge and the long legacy of
educational struggle and bureaucratic inertia in Seattle. These sedi-
mented patterns in the landscape have altered the foundation's preferred trajectory of reform in the Seattle Public Schools. But whereas in
many respects the initial funding priority can be said to have "failed"
to take root, in the larger schema we also show how this disconnection
led to the foundation's adaptation that has enabled (neoliberalizing)
educational reform to continue apace, but in a different form.

Choice-Based Education Reform and Its Institutional Infrastructure

In many ways, tactics of contemporary education reform are premised
upon a technique of government identified by Foucault as one of the

most basic concerns of modern governments: the management of risky populations.[3] Risk management has a particular genealogy as a technique of government reaching back to early biopolitical strategies to measure human life expectancy in actuarial terms. Within education the definition and management of risk takes on forms not typically cast in such terms, but that nevertheless hearken back both to biopolitical calculations of population management as well as to more neoliberal concerns with attracting capital. To address this risk, moral authority, expertise, and knowledge flow from philanthropic organizations to families and parents in a manner identified by Foucault (1991: 92) as a fundamental characteristic of liberal government: good government is drawn from, and modelled after, the morally laudable concern of fathers (and mothers) for their kin, while families are governed in a context that inculcates in them responsible, self-governing behaviour.

The idea of public education as a "risky" endeavour permeates American political culture. The 1983 report *A Nation at Risk*, while not explicitly advocating market-based school reform, codified and popularized the fear that American public education was failing to properly equip citizens to be economically competitive and culturally competent in an increasingly globalizing world. Indeed, much contemporary scholarship on governmentality in education analyses pedagogical and governance practices as risk-management strategies. Whether manifested in methods of student evaluation (Carlson, 2009) or teacher training (Stickney, 2009), such practices aim at identifying and mitigating potential risk by inculcating rational, self-entrepreneurial behaviour in governed subjects. These practices are endowed with moral authority by forecasting, as looming over the horizon, the consequences of not complying with them. For example, as Popkewitz (2008) points out, recent national-level reforms aimed at responsibilizing schools for student performance – such as the Bush-era No Child Left Behind (NCLB) and the Obama administration's Race to the Top (RttP) initiatives – imply the degenerate figure of the "no child" as the embodied failure of non-compliance to educational reforms. In addition, the devolving logic of school reform programs assigns not only authority over their execution, but responsibility for their success or failure, to administrators of school districts, mayors, principals, and ultimately families and students.

Indeed, education, like many areas of social governance, has been reorganized over the past several decades by the devolution of responsibility from larger scales of government (see Lippert in this volume;

Popkewitz, 2008). Neoliberal governance in education operates by delineating discrete fields of service and encouraging stakeholders to make their own opportunities to become involved (see Bloch, Popkewitz, Holmlund, & Moqvist, 2003; Peters, Besley, Olssen, Maurer, & Weber, 2009). This form of involvement is one that simultaneously reduces government responsibility from the management of structural relations even as it privileges an active citizenship based on the desire of individual actors to maximize their potential through making prudential calculations for themselves and their communities. Cruikshank (1999) claims further that this "will to empower" has the effect of motivating individuals, families, and communities to act on their own behalf while aligning their sense of agency with neoliberal policy goals.

The discourse of action-oriented empowerment has not remained purely abstract; indeed, it is manifested in multiple ways – from more negative definitions of freedom from regulation to more positive delineations of community pride in the pursuit of tackling a common problem. It is the latter manifestations that have most recently emerged in the arena of educational reform. The "problem" of public education became taken for granted since *A Nation at Risk* was published in 1983 (Berliner & Biddle, 1995; Blake, 2008), and the most recent variations on this theme involve the necessity to solve this problem with more rigorous forms of standardized assessment for students; teacher surveillance, training, and evaluation; and individualized responsibility for administrators and principals.

A growing number of philanthropic foundations have arisen over the past decade and a half to help provide the money and tools through which these conditions of rigour, expertise, and responsibility can be installed and the "problem" of public education resolved. Educational organizations, as well as schools, benefit from the links with these foundations, from which they draw a majority of their experimental funding, technical support, and policy expertise in setting the reforms into motion (Scott, 2009). Together, these advocacy and funding organizations act as conduits through which knowledge about the most prudential way to educate children flows from elite knowledge-making networks to communities and individual sites where school reform projects are actually carried out (Ravitch, 2010; Saltman, 2010).

For an illustration of how the reform ideology of self-styled experts heading philanthropic foundations flows through a complex network of intermediary educational organizations to meet up with a willing cadre of self-actualizing, entrepreneurial parents, students, and other activists, one need look no further than two popular cinematic portrayals of school reform: the 2010 documentary *Waiting for Superman* and the 2012 fictional (but billed as being based on events that are "making headlines daily") movie, *Won't Back Down*. In both films parents struggle against indifferent bureaucracies and self-interested teachers' unions in Harlem and Pittsburgh, respectively, to secure quality schooling for their children. In *Waiting* the mechanism for circumventing these roadblocks is the Harlem Children's Zone, where a limited number of charter school seats are distributed via lottery in a climactic and emotional final scene. In *Won't Back Down* the protagonists make use of so-called parent trigger laws that allow parents to undertake the radical transformation of public neighbourhood schools by firing staff or turning the school over to a charter operator.

Both films focus heavily on the role of administrative and parental activism, and both groups are depicted as drawing on expert knowledge about educational success in making their informed choices for educational reform. Behind the scenes, however, this apparently grassroots activism was heavily cultivated by philanthropic capital working through intermediary organizations. At the time *Superman* was being produced and the Harlem Children's Zone's high-stakes charter school lottery was entering public consciousness, for example, a campaign called "Flooding the Zone" was funded and promoted by the Success Charter Network and the Democrats for Education Reform with the explicit purpose of creating greater parental support in Harlem for charter school lotteries. These organizations drew funding from conservative foundations such as the New Schools Venture Fund. Similarly, the real-life inspiration for *Won't Back Down* can be found in a parent trigger law passed by the California legislature in 2010. The law was set into motion by a policy consultant for Green Dot Public Schools, a branch of charter schools in Los Angeles heavily funded by several pro-school reform organizations, including the Gates Foundation, the Broad Foundation, and the Walton Family Foundation.

These film-based examples reflect the vision and the push from above; they are examples of efforts to furnish expert knowledge about education and endow it with the moral authority of savvy parents and administrators, so that "prudent, calculating individuals and communities choosing "freely" and pursuing their own interests will contribute to the general interest as well" (Li, 2014: 37). These efforts of persuasion and provision are unquestionably uplifting and might seem irresistible.

But how often do these types of incentive and forms of conditioning actually lead to the preferred reform outcomes?

Unlike the happy endings of parents whose activation of their choice-making capabilities is validated through movie scripts and staged lottery victories, the "real-world" results often look quite different. The milieu in which administrators, parents, and other actors are nudged and recruited is not a flat plain, but reflects the memories of long periods of intervention and struggle that through time have been laid down in the educational landscape. These memories are part of broader inherited institutional topographies that often collide with the types of regulatory experiments initiated by the Gates Foundation and other philanthropists and trustees seeking to provide the conditions for educational reform.

Neoliberal governance of this variety relies on local organizational infrastructures as well as pro-responsibility and pro-school choice discourses tailored to local political cultures. For philanthropic organizations' educational ambitions to be realized, an assemblage must converge effectively in the historical and geographical specificities of a particular place. In Seattle we can see how this type of "optimal" milieu encounters various forms of place-based intransigence, such as foot-dragging, scepticism, weariness, inertia, and outright resistance among local actors and institutions. To what extent these actions were "merely obstructionist" (O'Malley, Weir, & Shearing, 1997), as opposed to deriving from locally understood and expressed rationalities such as pastoral power, which are constitutive of the social relations of governance (see Lippert, 2005: 9), was impossible to glean from the limited time we had with each informant. What *was* clear was that the foundation's preferred outcomes did not transpire, yet various kinds of adaptations and transformations in the overall project did take place, as we will see.

Learning from Small Learning Communities: The Gates Foundation Grant to Seattle Public Schools, 2000–2005

The movement to convert the large, general-education high schools in the United States into small learning communities (SLCs) has a complex history. National-level anxiety about the quality of US high schools, as exemplified by the *A Nation at Risk* report, led to a renewed interest in experimental educational models from the 1960s and 1970s that had been previously dismissed as insufficiently rigorous (Oxley, 1993; Wasley & Fine, 2000). They included experimental high schools known as "house systems," which grouped teachers and students into clusters that remained together through academic and extracurricular activities. Based on these past experiences and a handful of quantitative studies suggesting a link between smaller school size and higher student achievement, advocates began pushing for this style of learning, particularly in inner-city districts where dropout rates had reached alarming levels (David, 2008).

A growing need for resources and greater visibility on the part of SLC advocates was well matched by the priorities of the Bill and Melinda Gates Foundation in the early 2000s. Although the Gates Foundation is probably best known for its work abroad, particularly in sub-Saharan Africa, the foundation maintains an extensive portfolio of US-based projects in three major areas, including "College-Ready Education." Indeed, foundation leadership has been explicit in its commitment to fundamentally reform the American public school system, especially high schools. As Bill Gates declared in 2005:

> America's high schools are obsolete. By obsolete, I don't just mean that our high schools are broken, flawed and under-funded – though a case could be made for every one of those points. By obsolete, I mean that our high schools – even when they're working exactly as designed – cannot teach our kids what they need to know ... Today, only one third of our students graduate from high school ready for college, work and citizenship. The other two thirds, most of them low-income and minority students, are tracked into courses that won't ever get them ready for college or prepare them for a family wage job – no matter how well the students learn or the teachers teach. This isn't an accident or a flaw in the system; it is the system. (Gates Foundation, 2005)

As of 2001 the Gates Foundation had entrenched SLCs firmly within its overall strategy for education reform as a tactic designed to mitigate the risk of educational failure described by Gates and many of his contemporaries, cemented in part through "expert" assessments by outside consultants. Several reports supported by the foundation were written as part of an evaluation contract to assess reform programs within Washington State (e.g., Fouts & Associates, 2006), and outside the state (e.g., American Institutes for Research and SRI International, 2006). These reports highlighted "small, personalized learning environments" as one of the Gates Foundation's stated "Attributes of Effective High Schools." By breaking larger, more impersonal structures into smaller units the restructuring reform was intended to offer more frequent interactions among students and staff. At the same time, however, the smaller scale and more specific relations and interactions made it far easier to assess individualized student outcomes and attach these to specific teachers, administrators, and students themselves.

The possibility for more positive student-teacher interactions, in combination with the opportunity for employing more sophisticated metrics vis-à-vis student outcomes, was attractive to Gates Foundation staff. Indeed, even though one of the reports commissioned by the foundation noted that the research revealed mixed results in terms of student achievement in SLCs, program officers continued to allocate funding for this particular education reform initiative. Previous failures in SLCs were attributed to problems with implementation, and future successes were pinned directly on the "moral imperative" of administrators to change their practices and indeed their very beliefs in favour of these reforms (Fouts & Associates, 2006: 3). In other words, school leadership had failed to sufficiently exercise their moral authority in supporting and implementing the expert advice furnished by the Gates Foundation. For the programs to work, consultants indicated the necessity for board members and district administrators to not just take responsibility for the reforms, but to "*believe in* the type of educational experience the structural changes are meant to create" (emphasis added). A section titled "Lessons Learned: What Worked and What Did Not" illustrates this point clearly: "Successful conversion to SLCs requires a sustained commitment by the school district. Board members and district administrators must understand and support not only the structural changes being attempted but must also believe in the type of educational experience the structural changes are meant to create" (ibid.: 5).

In 2000 the Gates Foundation began one of its first large-scale interventions in fostering the creation of SLCs in Seattle Public Schools (SPS). This intervention began with a $25.9 million grant to the district as part of its Model District Initiative. According to a press release, the grant funds were earmarked to create a "model school district" on the basis of "improved teaching and learning, increased access to technology and stronger home and community partnerships" (Gates Foundation, 2000). The press release further noted that approximately 80 per cent of the grant monies were meant to be used in "every one of the District's ninety-seven schools for planning, technical assistance, staff training, and the purchase of technology resources that will transform schools into twenty-first-century learning organizations in line with the Foundation's Attributes of High Achievement Schools" (ibid.).

At the same time as the Model District Initiative grants were allocated to the SPS, the educators and administrators at Nathan Hale High School in north Seattle were independently searching for ways to improve performance at their academically struggling school. Looking outside the school for "best practices" from which to learn, they considered the SLC an attractive model to emulate. As one former Nathan Hale administrator said in an interview with us: "There were a core group of us who started learning about school reform ... We had a disproportionality committee that was looking at how can we better serve our students of colour, and then there was this core group of us that got really intrigued by school reform, and felt ... we could really ... change how we do schooling to better serve kids, not just tinker around the edges. And so ... we were able to ... make a design ... to better serve kids" (Rachel, former Nathan Hale administrator).

On the basis of this shared interest, Nathan Hale requested Initiative funds from SPS and, after implementing SLC reforms, produced improved short-term student results – as measured by the State of Washington Measure of Academic Process (MAP) standardized tests. It was at this point, however, that the SLC implementation process floundered. Principals at many other schools, especially those that were not academically struggling, resisted and even resented the district's insistence that they implement a project that was seen as targeting "underperforming" schools. In many cases teachers were left uninformed of the purpose of the grant monies and assumed – with their administrators' assent – that they were meant for technological upgrades and other direct classroom support materials rather than for educational reform. Our interviews indicated that many district-level administrators were unwilling to pressure school-level principals and teachers, and personnel turnover within the district meant that some administrators' initial enthusiasm and interest in SLC transformation was lost through attrition as well.

Although resistance and indifference to the SLC project was evident across the district, it was not uniform. In 2002 a former vice-principal who had helped implement Nathan Hale's SLC project was brought to Cleveland High School in south Seattle to foster its development there. Our interviews revealed that at first the transformation was profound and far-reaching, involving the break-up of the comprehensive high school and the creation of four themed learning academies. However, the sudden increase in administrative complexity accompanying the creation of four essentially separate schools led the school leadership to rapidly lose enthusiasm for the project. To help eliminate administrative overhead and regain trust, the number of academies was quickly reduced to three and then, following the elimination of the Technical Arts Academy, to just two. Interviewees revealed that fatigue and resistance at the teacher and administrative levels occurred in spite of measurable gains in student achievement within a short period of time. Similar difficulties occurred at other Seattle-area high schools in the original Model Districts Initiative; the initial enthusiasm of both teachers and administrators turned sour when they were confronted with monumental, complex, and largely unfunded logistical nightmares to resolve.

Thus, despite the initial success of the small learning academies at Nathan Hale and Cleveland High Schools, the Gates vision failed to take hold in Seattle. Resistance on the part of some schools to transform at all as well as logistical difficulties at the schools that did attempt the reforms convinced the school district by 2005 that the SLC initiative was not a project they wanted to pursue further, despite the fact that the term of the grant was not yet complete. The Gates Foundation responded by announcing that SPS as well as other school districts that had failed to produce results from the Model Districts Initiative would not be receiving grant renewals. This decision, as well as thinly veiled criticism of SPS, was expressed by a foundation official in the pages of the *Seattle Times*: "In choosing districts [for renewals of the Model Districts Initiative], the foundation selected those that 'have a really good record of improvement, have enjoyed stable and effective leadership, and had a really good plan going forward,' said Tom Vander Ark, the foundation's executive director for education giving. 'And none of those apply in Seattle'" (Shaw, 2005).

The Gates Foundation provided the conditions and incentives for greater individual responsibility, choice, and technological expertise and assistance through small-school transformation. But ultimately, the school district took the funding and the project in unanticipated directions, owing to the indifference or fatigue of individual administrators and teachers in sustaining the reforms. The reaction indicated above appears as a neoliberal disciplining of recalcitrant actors who refuse to comply with principles of self-entrepreneurship and inexorable calls to reform. Indeed, as one administrator who oversaw Cleveland's initial transformation recalls:

[handwritten annotations: "similar to idea from week 2 freedom + ..." / "governance of oneself is still salient and people take it away"]

And then the funding [for small schools transformation] came through from the Gates Foundation, and we were able to really do … it effectively there, but none of the other schools in the district were doing really anything along the lines of what they were supposed to. And I think ultimately what the Gates Foundation figured out – and they wrote a report about three or four years later … – was that if they didn't have the complete buy-in from the school district leadership and the willingness on the leadership's part to make sure that schools were doing what the grant money was intended to be doing, then it wasn't viable for them to give money on that scale to an institution like a school district. They needed guarantees, they needed accountability, they needed some expectation that what they were intending to have happen with that money was actually going to be done. (Bill, former Cleveland administrator)

The disciplinary aspect of the foundation's refusal to provide further funding for those districts that had failed to meet expectations is clear. But to see the attrition of the Gates Foundation's envisioned project as a simple failure of a neoliberal logic of private intervention in the public school system would be to miss a larger and more important consequence of the project. As the administrator quoted above noted, the Gates Foundation learned from the experience of giving money to a relatively large bureaucracy such as a district without having first ensured the compliance of relevant administrators and other staff. This policy learning and subsequent course correction was expressed neatly by the experience of Tyee High School – now called Global Connections Academy – located in Highline Public Schools, the district immediately to the south of Seattle.

In 2004, as the first round of Model District Initiatives was experiencing difficulty, the Tyee leadership expressed a desire to pursue an SLC transformation. The Gates Foundation, which by then had de-emphasized SLCs as an education reform strategy (Shaw, 2005), once again provided the money for this project. However, they made a key change in their lending policy: rather than grant the money directly to the school district, it was given to an organization called the Coalition for Essential Schools (CES). This intermediary organization now oversees SLC projects in a nationwide network of member schools.

In contrast to the unpredictable nature of public bureaucracies and potentially recalcitrant actors that the Gates Foundation had observed in the first round of its Model District Initiative grants, CES appeared to be an ideal partner for carrying out a project according to foundation

238 Katharyne Mitchell and Chris Lizotte

priorities. As related by an informant from the Gates Foundation, CES spends a great deal of time and resources vetting grant recipients based on mission alignment between the grantee and the foundation. CES, by its own declaration, provides assistance only to schools that have "embraced [its] Common Principles" (CES, 2015).

The choice to allocate funding through a third party represented a fundamental strategic change on the part of the Gates Foundation. Rather than recruit schools and districts directly through calls for grant proposals, the foundation opted to give money to an intermediary organization whose mission was allied with its own requirements for achieving measurable results from its investments. Interviewees explained that this move was consistent with a more general shift in granting guidelines that the organization was beginning to formulate. In particular, our Gates Foundation informant highlighted the benefits of working with such intermediaries. Unlike school districts, which the Gates Foundation regarded as "large, hard-to-move bureaucracies" subject to the whims of elected leadership, non-profit education organizations could be better counted on to maintain consistent priorities and aims in line with the foundation's own priorities.

Indeed, the shift from working with school districts to education reform organizations reflects in some ways an attempt to supplement the typical neoliberal impulse to push from above by harnessing potential agents of change from below: rather than having to take on the risky task of disciplining a potentially intransigent, complex entity such as a school district, the organization could call upon a flexible network of like-minded partners endowed with not only expert knowledge but also moral authority gained from their position as community-based organizations.

This change in granting practice also represented a fundamental geographic rescaling of the Gates Foundation's activities. Rather than distributing money directly to a locally delineated urban bureaucracy, with its inherent risks of political wrangling and administrative attrition, the foundation now gives its money to a national organization that draws on a shifting network of affiliated schools and centres. In doing so, the resources afforded by the Gates Foundation may be locally accessible, but they are vetted and filtered through a national network of best practices and knowledge sharing that sidesteps potential roadblocks to policy implementation. For Tyee High School in the Highline district, for example, the process of restructuring into multiple learning academies involved a circumvention of the typical chain of command within its own district. As one of the administrators who implemented the SLC transformation at Tyee recalled:

> It was an interesting process ... we had received a fairly large grant from the Coalition of Essential Schools, so we really had an external provider and when we got that part of the Memorandum of Understanding, it required us to break into small schools ... we really were lucky in that what they were asking us to do ... in the grant was what we wanted to do. But it gave us a different process with the [Highline] district in that the district signed a Memorandum of Understanding saying that we could do this. There was board approval at some point, but there wasn't any formal process. (Michael, former Tyee administrator)

What the Gates Foundation's shift in funding strategy ultimately reveals is the productive nature of a "failed" neoliberal policy. Because the foundation discovered that resistance or simply disinterest within large public bureaucracies is a potentially major roadblock to realizing its reform priorities, it altered its direction and its partnerships. In many cases this change circumvented school district authority in successive funding rounds, as the foundation began to work primarily with non-governmental intermediaries whenever possible. It should be noted, however, that the Gates Foundation has not abandoned the district altogether. Although the foundation redirected its major resources towards nationally networked intermediaries and away from local school districts, it continues to give small grants directly to Seattle Public Schools. Nevertheless, the foundation's overall strategy shift reflects a deepening entrenchment of neoliberal governmentality vis-à-vis the encouragement to bypass the public sector and rely on private sector organizations – whether for-profit or non-profit – in implementing community change.

Conclusion

Despite the funding incentives proffered for educational transformation, the neoliberal rationalities of constant entrepreneurial reform were unsustainable in Seattle's public schools. Ultimately, they failed to connect with administrators and teachers, who were either disinterested from the outset or became weary with the ongoing logistical challenges associated with the reform's implementation. In other cases administrators and teachers took the grant funding with no intention of carrying out the reforms – spending the money on needed school supplies instead. However, although these types of disconnection might be analysed as a ground-up failure with respect to the inculcation of rational choice making and entrepreneurial behaviour in this specific local milieu, it can also be seen as a learning moment – one that ultimately led to a recalibration of the Gates Foundation's granting guidelines and to new, more effective and sustainable modes of neoliberal governance operating at different levels.

What an ethnographic approach to the processes of neoliberalization can thus show us is the often complex, *both/and* nature of neoliberal governmentality. It can manifest the attempts to know and regulate the processes associated with populations, "to know that which was to be governed and to govern in light of that knowledge" (Rose et al., 2006: 87) – in this case those systems and regulations associated with education – and it can show the various disarticulations that make this governing fail. But even more, grounded ethnographic methods *in concert with* a theoretical awareness of the path-dependent patterning of regulatory landscapes through time can also reveal new phases, outgrowths, and spatial scales of neoliberalism that may result from these failures.

ACKNOWLEDGMENT

Parts of this chapter were first published in a 2014 issue of *Foucault Studies*, 18: 66–89.

NOTES

1 The most prominent of these is Teach for America.
2 A charter school is a publicly funded but privately managed school
   that, depending on local state law, has greater flexibility in personnel
   management and pedagogic aspects than traditional public schools have.
3 Portions of this section have appeared in Mitchell & Lizotte (2014).

REFERENCES

American Institutes for Research and SRI International (2006). *Evaluation of the Bill and Melinda Gates Foundation's High School Grants Initiative: 2001–2005*

Final Report. Available at https://docs.gatesfoundation.org/Documents/
year4evaluationairsri.pdf. Accessed 4 April 2015.

Atia, M. (2013). *Building a house in heaven: Pious neoliberalism and Islamic charity in Egypt*. Minneapolis: University of Minnesota Press. http://dx.doi.org/10.5749/minnesota/9780816689156.001.0001

Berliner, D., & Biddle, B. (1995). *The manufactured crisis: Myths, fraud, and the attack on America's public schools.* Reading, MA: Addison-Wesley.

Bishop, M., & Green, M. (2008). *Philanthrocapitalism: How the rich can save the world*. London: Bloomsbury Press.

Blake, S. (2008). A nation at risk and the blind men. *Phi Delta Kappan, 89*(8), 601–2. http://dx.doi.org/10.1177/003172170808900814

Bloch, M., Popkewitz, T., Holmlund, K., & Moqvist, I. (2003). *Governing children, families and education: Restructuring the welfare state*. New York: Palgrave Macmillan.

Brady, M. (2011). Researching governmentalities through ethnography: The case of Australian welfare reforms and programs for single parents. *Critical Policy Studies, 5*(3), 264–82. http://dx.doi.org/10.1080/19460171.2011.606300

Brady, M. (2014). Ethnographies of neoliberal governmentalities: From the neoliberal apparatus to neoliberalism and governmental assemblages. *Foucault Studies, 18*(1), 11–33.

Brenner, N., Peck, J., & Theodore, N. (2010). Variegated neoliberalization: Geographies, modalities, pathways. *Global Networks, 10*(2), 182–222. http://dx.doi.org/10.1111/j.1471-0374.2009.00277.x

Carlson, D.L. (2009). Producing entrepreneurial subjects: Neoliberal rationalities and portfolio assessment. In M. Peters, A. Besley, M. Olssen, S. Maurer, & S. Weber (Eds), *Governmentality studies in education* (257–70). Boston: Sense.

CES (2015). Mission statement. Available at http://essentialschools.org/what-we-do/. Accessed 1 April 2015.

Cohen, D., & Lizotte, C. (2015). Teaching the market: Fostering consent to education markets in the United States. *Environment & Planning A, 47*(9), 1824–41.

Collier, S. (2012). Neoliberalism as big leviathan, or ...? A response to Wacquant and Hilgers. *Social Anthropology, 20*(2), 186–95. http://dx.doi.org/10.1111/j.1469-8676.2012.00195.x

Cruikshank, B. (1999). *The will to empower: Democratic citizens and other subjects*. Ithaca, NY: Cornell University Press.

David, J. (2008). What research says about small learning communities. *Educational Leadership, 65*(8), 84–5.

Foucault, M. (1991). Governmentality. In C. Burchell, C. Gordon, P. Miller, &
M. Foucault (Eds), *The Foucault effect: Studies in governmentality. With two
lectures by and an interview with Michel Foucault* (87–104). London: Harvester
Wheatsheaf.

Foucault, M. (2008). *The birth of biopolitics: Lectures at the Collège de
France, 1978–1979.* New York: Palgrave Macmillan. http://dx.doi.
org/10.1057/9780230594180

Fouts & Associates. (2006). *Leading the conversion process: Lessons learned and
recommendations for converting to small learning communities.* Available at http://
www.learningace.com/doc/5456744/1bd5869ccd5271f71e03b9662910dd04/
leading-the-conversion-process-10-6-06. Accessed 4 April 2015.

Gates Foundation (2000). The Bill and Melinda Gates Foundation commits
$25.9 million to the Alliance for Education and Seattle Public Schools
to help all children achieve in the classroom. Available at http://www.
gatesfoundation.org/Media-Centre/Press-Releases/2000/03/Alliance-for-
Education-and-the-Seattle-Public-Schools. Accessed 4 April 2015.

Gates Foundation (2005) Bill Gates – National summit on high schools. Available
at http://www.gatesfoundation.org/media-center/speeches/2005/02/bill-
gates-2005-national-education-summit. Accessed 4 April 2015.

Hart, G. (2002). *Disabling globalization: Places of power in post-apartheid South
Africa.* Berkeley: University of California Press.

Harvey, D. (2006). *Spaces of global capitalism: A theory of uneven geographical
development.* New York: Verso.

Hilgers, M. (2010). The three anthropological approaches to neoliberalism.
International Social Science Journal, 61(202), 351–64. http://dx.doi.
org/10.1111/j.1468-2451.2011.01776.x

Hunt, A. (2004). Getting Marx and Foucault into bed together! *Journal
of Law and Society, 31*(4), 592–609. http://dx.doi.org/10.1111/j.1467-
6478.2004.00305.x

Hunt, A., & Wickham, G. (1994). *Foucault and law: Towards a sociology of law as
governance.* London: Pluto Press.

Kovacs, P. (2011). *The Gates Foundation and the future of U.S. "public" schools.*
New York: Routledge.

Lagemann, E. (1989). *The politics of knowledge: The Carnegie Corporation,
philanthropy, and public policy.* Middletown, CT: Wesleyan University Press.

Leitner, H., Peck, J., & Sheppard, E. (2007). *Contesting neoliberalism: Urban
frontiers.* New York: Guilford Press.

Li, T. (2007). *The will to improve: Governmentality, development, and the
practice of politics.* Durham, NC: Duke University Press. http://dx.doi.
org/10.1215/9780822389781

Li, T. (2014). Fixing non-market subjects: Governing land and population in the global south. *Foucault Studies, 18*(1), 34–48.

Lipman, P. (2011). *The new political economy of urban education: Neoliberalism, race, and the right to the city*. New York: Routledge.

Lippert, R. (2005). *Sanctuary, sovereignty, sacrifice: Canadian sanctuary incidents, power, and law*. Vancouver: UBC Press.

Lippert, R., & Stenson, K. (2007). Urban governance and legality from below. *Canadian Journal of Law and Society, 22*(2), 1–4. http://dx.doi.org/10.1017/S0829320100009315

McCann, E. (2011). Urban policy mobilities and global circuits of knowledge: Towards a research agenda. *Annals of the Association of American Geographers, 101*(1), 107–30. http://dx.doi.org/10.1080/00045608.2010.520219

McGoey, L. (2012). Philanthrocapitalism and its critics. *Poetics, 40*(2), 185–99. http://dx.doi.org/10.1016/j.poetic.2012.02.006

Mitchell, K., & Lizotte, C. (2014). The grassroots and the gift: Moral authority, American philanthropy, and activism in education. *Foucault Studies, 18*, 66–89.

Moore, D. (2005). *Suffering for territory: Race, place, and power in Zimbabwe*. Durham, NC: Duke University Press. http://dx.doi.org/10.1215/9780822387329

O'Malley, P., Weir, L., & Shearing, C. (1997). Governmentality, criticism, politics. *Economy and Society, 26*(4), 501–17. http://dx.doi.org/10.1080/03085149700000026

Ong, A., & Collier, S. (2004). *Global assemblages: Technology, politics, and ethics as anthropological problems*. Oxford: Wiley-Blackwell.

Oxley, D. (1993). *Organizing schools into smaller units: A planning guide*. Available at http://files.eric.ed.gov/fulltext/ED364948.pdf. Accessed 4 April 2015.

Peck, J. (2010). *Constructions of neoliberal reason*. Oxford: Oxford University Press. http://dx.doi.org/10.1093/acprof:oso/9780199580576.001.0001

Peck, J., Theodore, N., & Brenner, N. (2010). Postneoliberalism and its malcontents. *Antipode, 41*(S1), 94–116. http://dx.doi.org/10.1111/j.1467-8330.2009.00718.x

Peters, M., Besley, A., Olssen, M., Maurer, S., & Weber, S. (2009). *Governmentality studies in education*. Boston: Sense.

Popkewitz, T. (2008). *Cosmopolitanism and the age of school reform: Science, education, and making society by making the child*. New York: Routledge.

Ravitch, D. (2010). *The death and life of the great American school system: How testing and choice are undermining education*. New York: Basic Books.

Ravitch, D. (2013). *Reign of error: The hoax of the privatization movement and the danger to America's public schools*. New York: Knopf.

Reckhow, S. (2012). *Follow the money: How foundation dollars change public school politics.* Oxford: Oxford University Press. http://dx.doi.org/10.1093/acprof: oso/9780199937738.001.0001

Rose, N., & Miller, P. (1992). Political power beyond the state: Problematics of government. *British Journal of Sociology, 43*(2), 173–205. http://dx.doi.org/10.2307/591464

Rose, N., O'Malley, P., & Valverde, M. (2006). Governmentality. *Annual Review of Law and Social Science, 2*(1), 83–104. http://dx.doi.org/10.1146/annurev.lawsocsci.2.081805.105900

Roy, A. (2012a). Subjects of risk: Technologies of gender in the making of millennial modernity. *Public Culture, 24*(1), 131–55. http://dx.doi.org/10.1215/08992363-1498001

Roy, A. (2012b). Ethnographic circulations: Space-time relations in the worlds of poverty management. *Environment & Planning A, 44*(1), 31–41. http://dx.doi.org/10.1068/a44180

Saltman, K. (2010). *The gift of education: Public education and venture philanthropy.* New York: Palgrave Macmillan. http://dx.doi.org/10.1057/9780230105768

Scott, J. (2009). The politics of venture philanthropy in charter school policy and advocacy. *Educational Policy, 23*(1), 106–36. http://dx.doi.org/10.1177/0895904808328531

Shaw L. (2005, 20 October). Gates Foundation exec pans Seattle school district. *Seattle Times.* http://www.seattletimes.com/seattle-news/gates-foundation-exec-pans-seattle-school-district/.

Smith, N. (2008). *Uneven development: Nature, capital, and the production of space.* Atlanta: University of Georgia Press.

Stickney, J. (2009). Casting teachers into education reforms and regimes of inspection: Resistance to normalization through self-governance. In M. Peters, A. Besley, M. Olssen, S. Maurer, & S. Weber (Eds), *Governmentality studies in education* (235–56). Boston: Sense.

Wacquant, L. (2012). Three steps to a historical anthropology of actually existing neoliberalism. *Social Anthropology, 20*(1), 66–79. http://dx.doi.org/10.1111/j.1469-8676.2011.00189.x

Wasley, P., & Fine, M. (2000). *Small schools and the issue of scale.* New York: Bank Street College of Education.

Zunz, O. (2011). *Philanthropy in America: A history.* Princeton: Princeton University Press.

11 Exploring the Complexity and Contradictions of Poverty Governance: The Case of Payday Lending in Australia

GREG MARSTON

My aim in this chapter is to explore the growth of fringe financial lenders within the broader context of institutional change in poverty governance in Australia, in order to reveal theoretical and practical insights into the hybrid mix of welfare and market rationalities within contemporary western societies. Similar to other liberal welfare states (Esping-Andersen, 1990) such as the United States, the United Kingdom, Canada, and New Zealand, Australia has a mixed economy of welfare where the non-profit and the private sectors play a substantial role in the delivery of social services, alongside government providers. While there have been numerous studies of the implications of contract and privatized welfare within this group of countries, there has been much less research on the private businesses that are playing an implicit and often unacknowledged but significant role in poverty management through the provision of high-cost, short-term credit to citizens in geographical areas where there is a high concentration of socio-economic disadvantage (Karger, 2005). Ironically, the fringe lenders who provide credit to these citizens do not see themselves as providing a welfare service. Given these developments, it is important to consider some analytical inroads into the human and spatial configurations that are playing out on a day-to-day basis in such local urban environments.

This chapter seeks to bring political economy and analytics of governmentality approaches together to examine these configurations. Thus, rather than only undertaking a political economy analysis, which focuses on demand and supply factors that have led to the increase in fringe lending, or applying an analytics of governmentality, which would seek to understand how payday lending has been problematized and governed through specific technologies and subjectivities, this chapter aims to conduct both analyses alongside each other and to hold them in tension. Drawing on empirical research, which involved analysis of policy and legal reforms, and interviews with payday lenders, borrowers, regulators, and consumer credit advocates conducted from 2010 and 2012 in three Australian states, this chapter argues that the fringe lending sector has expanded in response to increasing economic insecurity and that Australian payday borrowers experience competing discourses of paternalism and free choice when accessing high-cost credit. Such attention to context and experience accords with the aim of this volume to engage with everyday practice to understand the social realm while drawing on an analytics of governmentality. The chapter proceeds in three parts. First, I sketch the background of this expanding fringe lending sector, particularly the impact of economic insecurity brought about by labour market changes, rising living costs, and inadequate state-provided income support payments. Against this background I theorize questions of agency and identity in the fringe lending sector in the second part of the chapter. The third part of the chapter uses an analytics of governmentality perspective to highlight the subjectivities of payday borrowers and lenders and how lenders seek to "conduct the conduct" of borrowers.

A Brief Overview of Payday Lending

The growth of payday lenders in Australia can be traced back to the deregulation of the Australian banking sector in the early 1980s (Scutella & Sheehan, 2006). In Australia mainstream banks have increasingly discouraged low-income consumers from using their credit services, both by raising basic transaction costs and by removing financial products tailored to their needs, such as small personal loans (Leyshon & Thrift, 1997: 226). Howell (2005) argues that many borrowers simply assume that mainstream borrowers will not assist them, and they are wary of credit cards and do not trust larger institutions and banks. Fringe lenders have capitalized on the negative perception of banks and typically provide a quick and easy service and work hard to make customers feel welcome (Marston & Shevellar, 2010a; Howell, Wilson, & Davidson, 2008).

Consumer advocates, such as financial counsellors, tend to reframe this "friendly service" orientation of the payday lender as "predatory lending," a term used to describe how some sections of this growing credit industry entice and induce a borrower to take out a loan with high fees and a high interest rate in a less than transparent way (Wilson,

2004). On the demand side there is growing income equality, rising living costs (particularly housing), and insecure and entrenched pockets of
long-term unemployment in Australia, particularly for young people.
All of these factors can create household demand for high-cost, short-
term credit to make ends meet. Studies into the typical payday borrower
show that there is some variation between countries in terms of borrowers' characteristics, but one common characteristic is living on an annual
income towards the bottom of the income scale (Elliehausen, 2009).

Another common characteristic of the payday-lending market is
repeat borrowing. King & Parrish (2007) argue that the payday-lending
business model actually *depends* on trapping borrowers in loans. It is
these kinds of pattern that leads critics to refer to payday lending as the
credit market's equivalent of crack cocaine: a highly addictive source
of easy money that hooks the unwary consumer into a cycle of debt
(Banks, Marston, Russell, & Karger, 2012; Stegman, 2007).

In the mixed economy of credit there are various community-based
microfinance alternatives that are attempting to break this pattern of
high-cost borrowing, although their impact has been muted, owing to
the fact they do not operate on the same scale as payday lenders and
are not always meeting the same need (see Chant Link and Associates
2004). While it is important to acknowledge the role of these lenders,
many of the community-based microfinance schemes do not meet the
same credit need as payday lenders, since the microfinance schemes
tend to tie the loans to small capital purchases, rather than provide
loans for recurrent items (such as utility bills, rent, and food) (Marston
& Shevellar, 2010a).

The mix of non-profits and for-profits in this particular space highlights the hybrid nature of poverty governance in contemporary western societies, a point that can get lost in totalizing accounts that locate
recent developments within processes of commodification associated
with the influence of neoliberal rationalities. The institutional and
policy measures in poverty governance are well known: from creating internal markets, market testing, contracting out, privatization,
and encouraging private pensions and superannuation, to restricting
access to higher payments such as pensions, lowering the amounts pay-
able, and generally ramping up coercive workfare programs to force
labour in poorly paid (or even unpaid) work (Jessop, 1999; Lavalette
& Mooney, 1999; Schram, Soss, Houser, & Fording, 2010). Within the
payday-lending literature neoliberalism therefore tends to be represented as a political ideology and program of reform that has made

payday lending a "necessary evil" and is closely associated with the decline of the coverage of state-based social welfare programs in the latter part of the twentieth century (Karger, 2005; Rivlin, 2011). While this narrative about recent social changes in light of neoliberal reforms is compelling, it is important not to ignore the complexity of cultural and political differences between and within countries (Brady, 2014).

The specific political and economic context of lending has an important bearing on how people engage in the mixed economy of credit. Citizens are exercising constrained choices based on the location of different providers, cost, eligibility, and knowledge of available options. This economic and geographical calculus is also influenced by the emotional dimension, as in the feelings of shame or guilt that may be associated with asking parents or friends for a short-term loan. While the financial cost may be low, the emotional cost associated with taking the loan may be very high. The individual citizen must therefore weigh the different transaction costs when seeking access to credit. Although money is an object – an inert thing – it is also subjective, with affective meanings that influence people's attitudes and behaviour (Banks et al., 2012; Martin, 2002).

The moral codes governing decisions about debt also change over time and across space. Reflecting on the media commentary about increasing household and national debt led Canadian writer Margaret Atwood to conclude that a fallout from the global financial crisis is that "we seem to be entering a period in which debt has passed through its most recent harmless and fashionable period, and is reverting to being sinful" (2008: 32). While the political debate about the appropriate personal and policy response to greater levels of household debt and failing financial markets continues at national and international levels, there has been much less focus on knowledge about the competing pressures and new configurations of economic and social relations between governments, low-income citizens, and markets at the local urban level. To appreciate the multidimensional nature of this cultural and material space requires theoretical approaches that are capable of working at multiple levels of analysis and responding adequately to contradictions and complexity.

Theorizing the Place of the Payday Lender in Poverty Survival

In this section I want to theorize poverty survival and the role of payday lenders, using the insights from various modes of political economy augmented by an analytics of governmentality to highlight the structural and subjective factors shaping financial decision making in low-income households. In classical political economy Marx sought to identify the structural forces associated with the rise of capitalism. Though clearly influenced by Marx, Foucault examines history in terms of contingencies and events, rather than a series of specific epochs (Walters, 2006). There is a considerable body of scholarship that seeks, in the apt words of Hunt (2004), to get Marx and Foucault "into bed together"; however, there is still a deep suspicion within social policy and social welfare studies that not much good will come of such a partnership (Lavalette & Pratt, 2006; Marston & McDonald, 2009). The dominance of the normative stance in social policy research of identifying the conditions of the "good society" tends to preclude a governmental analytics that serves a different set of moral forces – those concerned with destabilizing assumptions (Dean, 2010).

The combination of these modes of theorizing social change are necessary to develop an approach that is capable of looking at both the *why* questions (why did things happen as they did?) and the *how* questions (by what means and specific actions?) (Hunt, 2004: 604). Springer (2012) demonstrates in his analysis of neoliberal reason how post-structuralism and culturally informed political economy can be merged by focusing on actual practices, both discursive and material. This involves recognizing the importance of both critical perspectives without privileging either.

We can analyse poverty and poverty governance, which is the focus of this chapter, as (1) a material phenomenon caused by a mix of structural and personal factors (Spicker, 2007); (2) a discursive construct, where there are competing moral definitions of absolute and relative poverty (Lister, 2004); and (3) an object of governmental technologies to teach citizens new moral and material habits (Schram, 2006). In this chapter the growth in fringe lending is analysed at the macro level as partly resulting from material structural changes, including inadequate welfare and available paid work that leads to insecure household income (Standing, 2011), and at the micro level as an example of poverty governance where people are making constrained choices, influenced by competing moral positions about access to credit and the accumulation of debt.

The term "shadow welfare state" helps to capture some less regulated modes of poverty survival where local-level market and voluntary actors fill the gap left by patchy state-funded welfare services and has been used to examine how the private sector involvement in health and welfare is supported by government policy (see Gottschalk, 2000). Fairbanks (2009: 273) has developed the term to refer to "the many informal assemblages of collective responsibility and self-help operating in the tradition of voluntarism in the post-welfare age. While apparently decoupled from the state apparatus at first glance, myriad configurations of the shadow welfare state have emerged to forge complex partnerships with state systems – primarily in response to devolutionary trends."

The devolutionary trends referred to in Fairbanks's definition reflect the proposition that the state is either outsourcing or withdrawing from direct service provision in key areas of social policy, while at the same time remaining involved as a regulator and auditor, and it is thus "governing at a distance" (Rose, 1999). We must be cautious about overextending the metaphor of "state rollback," given that most western liberal welfare states have historically had a diverse mixed economy of welfare (Harris & McDonald, 2000) and are proving more resilient to spending cuts than many suggested (Castles, 2004; Pierson, 2001; Whiteford, 2011). Much of the transformation in western liberal welfare states is not about declining state finances, but is about how services are being provided and the shifting discourse around risk and responsibilization.

Some commentators contend that the growth in privatized solutions to poverty is entirely predictable in contemporary liberal welfare states, where collective risk pooling is being eroded by cuts to cash and in-kind benefits and various forms of self-management and individualization of welfare and well-being (Dean & Melrose, 1996; Hacker, 2006; Jordan & Travers, 1998). As Karger (2005: 15) argues, cost containment in state welfare reform initiatives and "fringe lending" are like two peas in a pod: "In some measure, the fringe economy represents a privatized and expensive de facto welfare state since it offers former recipients emergency cash services no longer furnished by the government."

As a consequence of these institutional shifts there are many small businesses, such as payday lenders, that have quickly established themselves in the urban fringes of capital cities where poverty has become spatially concentrated. In many US cities, cheque cashiers, payday lenders, and pawn shops now lease shopfronts that were previously tenanted by retail outlets and local grocery stores (Karger, 2005). A similar spatial pattern emerges in Australia; however, the presence of the bureaucratic welfare state often coexists in the same geographical space as the private payday-lending outlets in this country. In outer urban suburbs of major capital cities in Australia, for example, Centrelink offices (the statutory income support agency) are often on the same street as the payday lenders, and these lending businesses openly advertise to people on social security benefits through billboards and websites that have marketing slogans such as "Pensioners Welcome."

Moreover, the "friendly face" of the payday lender that greets citizens when they walk in off the street stands in stark contrast to the indifference that many people encounter in Centrelink offices in Australia (De Parle, 2004; Murphy, Murray, Chalmers, Martin, & Marston, 2011). This contrast between the indifferent welfare state and the friendly face of the payday lender affirms the analytical importance of emotion and recognition in understanding individual decision making about access to credit in low-income households. The spatial proximity between the payday-lending outlets and the statutory welfare office is also symbolic of the "economic interdependence" between public welfare and the "shadow welfare state." Government-provided income support payments to social security recipients are used to secure loans from private lenders and/or to repay existing loans. In this respect the state and the market are not neatly decoupled in poverty governance; hence the aptness of the shadow welfare state metaphor.

While recent developments in political economy help illuminate the dynamics of institutional change that promote blurred boundaries between public and private welfare (see Streeck & Thelen, 2005), there continues to be a tendency in this mode of theorizing to overdetermine the subjugation of the state to the market (Clarke, 2006). The conventional analysis of rolled-out neoliberalism in political economy can miss the ways this political project is realized and challenged within different national contexts. Consequently, we end up with a somewhat thin account of social change because of a reliance on a "top-down" analysis. As Wacquant (2009: 306) argues: "We need to reach beyond this economic nucleus and elaborate a thicker notion that identifies the institutional machinery and symbolic frames through which neoliberal tenets are being realized." This means we need to analyse not only the modes in which neoliberal ideas and practices are rolled out, but also the forms of resistance, adaptation, and hybridization that are encountered in practice (see Brady and Lippert in this volume). Resistance may take unanticipated forms in reshaping the market offensive. The position taken in this chapter is that neoliberalism is a not a perfectly integrated and contradictory free system of globalism (Peck, 2010: 28).

There is always room for considerable variability across the global political economy, particularly when the local scale is brought into the analytical frame.

It is at this level of analysis that a governmentality framework becomes productive, particularly a governmental analysis able to account for not only how subjects are constituted, but how subjects respond, adapt, and modify to governmental technologies aimed at governing subpopulations and individual citizens. Such an analytical starting point also moves beyond some governmentality studies that simply "read off" subjectivities from official texts and documents (Brady, 2014). As Clarke (2006: 90) argues: "Governmental technologies – and their conceptions – represent specific attempts at mapping, but they have to negotiate both pre-existing and emergent mappings." This account of the "conduct of conduct" offers a richer view of the social landscape, one that acknowledges the many actors outside the state (such as payday lenders) that are taking an increasingly active role in poverty governance. In this respect Foucault's analytics of governmentality offers a less state-centric concept of government compared with that offered by traditional political economy (1991: 87).

In terms of social policy analysis, a governmentality framework helps to problematize the narrative of welfare state retrenchment and the romanticization of the agency of the working poor in late capitalist societies. As Fairbanks (2009: 20) argues: "A governmentality approach has the potential to address the traditional pitfalls of moralizing or pathologizing the poor ... precisely by opening us up to the ways in which liberal governance often operating through the apparatus of social policy, presupposes rather than annuls the agency of poor subjects." This analytical frame allows us to fully consider the modes of governing and moral dilemmas in welfare regimes that are characterized by a mix of formal state organizations, market actors, and the informal sector at the local level of poverty management. Such attention to actual practices of governing accords with Dean's (2010: 12) conception of an analytics of governmentality where the emphasis is on identifying practices of government. In these terms the "welfare state" is understood less as a concrete set of institutions and more as a way of viewing institutions, practices, and personnel or organizing them in relation to a specific ideal of government, such as "small government" (ibid.: 43). In a devolved mode of welfare governance there are many actors with varying degrees of formal authorization that are playing a significant role in poverty governance.

In anglophone countries the discipline of markets is considered both morally and economically superior to dependence of low-income groups on the state (Schram, 2006). Moreover, a social policy solution that is geared towards legal regulation of financial lending depoliticizes poverty – as the focus becomes "getting the balance right" between consumer protection and reducing "red tape" for fringe credit providers. The result can obfuscate power relations between classes and groups in society. Regulating money lending through law reform legitimizes the role of the sector and helps promote a dominant rationality of poverty governance in countries like Australia, that is, to manage low-income populations and transform them into cooperative subjects of the market and polity (Soss, Fording, & Schram, 2011: 2). As Ong (2006) observes, neoliberalism is an energetic, ongoing process of recasting social problems as non-political and non-ideological matters that require technical solutions. And it is a process that is accomplished through countless individualized micro-management processes, where citizens are expected to transform themselves, with the guidance of others, into responsible citizens through the implementation of a realistic household budget or a personal résumé that enhances employability. The state, civil society, non-profits, and market actors all play a role here, with differing degrees of legitimacy. The place of payday lenders in poverty governance is highly contested. Many consumer advocates position them as immoral actors, being driven only by a concern for profit, with little or no interest in the welfare of the borrower (Gillam, 2010). What I want to suggest in the following analysis is that the norms, expectations, and ideas of obligation between lenders and borrowers in payday lending are more complex.

The Australian Case: Poverty, Power Relations, and Subjectivity

It is difficult to research the fringe credit industry because it is broad, multifaceted, and forever changing its products in response to new government regulations. Australia, like other English-speaking countries, has been engaged in an intense policy debate about how to regulate the growing fringe credit industry (Gillam, 2010). In the national policy debate in Australia there are competing constructions of where the problem lies and thus what the course of action should be. Lenders and their representatives construct the policy problem as a need to "clean up" the industry. "Light" market regulation, they argue, is the solution (NFSF, 2010). In contrast, welfare and consumer activists consider the existence of the payday industry a problem that needs to be addressed through stringent interest rate caps or the abolishment of payday lending altogether. The voices that are typically missing from this polarized policy debate and high-level policy discussions are those of the borrowers.

Part of the motivation for the ethnography that was conducted in 2010 and 2012 was to capture some of these voices to see how much they reflected the representation of consumer advocates and industry representatives. Lenders were interviewed to see what rationalities they drew on in relation to their purpose and function, while borrowers were interviewed to provide insights into how they constructed the process of securing a loan. The study was conducted in the states of Queensland, Victoria, and New South Wales, Australia, and involved a total of 25 street surveys with borrowers, 122 in-depth interviews with borrowers, 20 interviews with lenders, 45 interviews with consumer advocates, and analysis of policy and legal reforms. While there have been many studies of the legal regulatory environment that governs these lending practices (Wilson, 2004; Consumer Action Law Centre, 2013), few of them have sought to reveal the views of those who inhabit the field, which can unsettle assumptions about villains and victims in policy and practice debates.

The Perspective of Payday Lenders

While playing a direct material role in poverty management, what was clear from our interviews with industry representatives and individual lenders was that they were reluctant to associate their service or product with "poverty relief." They preferred to use a neoclassical economic rationality of simple supply and demand, rather than a social or moral rationality to justify their service and product. In part, this may be a strategic response to the way in which academics and media commentators have portrayed the industry as having a "predatory relationship" with low-income individuals and households (Karger, 2005). As a result, lenders have been keen to distance themselves from an image of "preying on the poor."

What is perhaps surprising, then, is how many of the lenders involved in the study acknowledged the need to take an interest in the welfare of the person and their capacity to repay. At the same time as lenders were keen to differentiate themselves from an essential service provided by the state, they were keen to point out that they offer a product, not a social service, as the following quotation from a small lender makes clear: "One of the first things that it's important to look at from a lender's point of view is, this is a business and it has to be managed as a business. The first thing is, in this case, it is a small retail business. The product just happens to be money" (payday lender, Gold Coast, Queensland). Some lenders, however, overlaid this market discourse with an explicit recognition of the social context of financial decision making for individual borrowers: "Most of the people I speak to are on Social Security payments, not because that's the way we want it but they're the ones who come to us because they can't go anywhere else. I am looking at the payday-lending area, and once again I don't see anything wrong with that, because it fills a need" (payday lender, Melbourne, Australia; all quotes in this paragraph are from Marston & Shevellar, 2010a).

Meeting needs through responsible lending is a dominant discourse not only in payday lending but also in the wider financial service sector, particularly in light of the sub-prime mortgage crisis in the United States. The peak body representing payday lenders in Australia, the National Financial Services Federation (NFSF), has drafted a code of practice for its members. The NFSF Code of Ethics includes statements such as "We believe the interests and needs of our business are best served by providing adequate protection for the interests and needs of our customers" and "all representations shall be truthful, without exaggeration, concealment or omission"; finally, there is a claim to professional recognition in the statement: "Our staff regard the provision of financial services as an important and responsible profession" (NFSF, 2010). We can read this text as an attempt at reframing a negative organizational identity in the mixed economy of credit, particularly in the context of a public moral discourse that has generally positioned lenders as irresponsible, deceptive, and acting with little concern for the material hardship faced by those that access their services (Banks et al., 2012).

Small fringe lenders' construction of their role was largely consistent with the principles expressed in the NFSF Code of Ethics. One lender interviewed for the study went as far as to say that he always tells new customers that his message to borrowers is to be a "reluctant lender": "'I don't want to lend to you, I don't care if I don't do a loan to you, but if you've got a real need I will give you a loan and I will tell you what it will cost.' It's quite fascinating to have that attitude towards lending; I know there's a lot of lenders out there that are pretty 'sharky,' but the same with lots of other industries, you can go a hire a bloody builder and he's a shark or get your car repaired and they over-service it" (payday lender, Brisbane, Australia; Marston & Shevellar, 2010a).

This differentiation between "sharky" and ethical was often expressed in terms of the difference in a payday-lending operator's size. The smaller, single-site operations regularly contrasted themselves with the multi-site operators and large franchises of the fringe lending sector, who were positioned as less ethical in regard to "responsible lending." Another differentiation made between ethical and unethical lending practice concerned whether lenders extended too much credit without obtaining sufficient knowledge about the purpose of the loan. Lenders thus distinguished between good and bad lending: "However, to me I want to lend for a genuine reason, not just top up someone's pocket 'til pay day sort of thing. I think payday lending has a place, and the way some of the payday lenders operate where they take such a large chunk of the person's pay when it does come in, I think that's bad too" (payday lender, Brisbane).

An effort by lenders to discern legitimate from non-legitimate reasons for the loan challenges the dominant construction (in policy debates) of lenders as people who ask too few questions and are too quick to lend money to people who have limited capacity to repay the loan. However, borrowers themselves did not necessarily support lenders' claims, some seeing it as too easy to get money. Making assessments about whether to lend or not was sometimes based on both evidence of capacity to pay and moral judgments about financial management skills, linked to a neoliberal discourse of market-based self-reliance. As one lender indicated: "I mean I have people here who haven't been able to afford their bread and milk and stuff and they've been on Centrelink benefits. But my suspicion is that they just don't manage their money well, so it's those sorts of people I don't help. I just send them on to get food vouchers" (payday lender, Melbourne, Australia).

Lenders used specific moral discourses when discussing borrowers, which are not that dissimilar to traditional welfarist discourses of "deserving" and "undeserving poor" (Piven & Cloward, 1971). Borrowers were assigned to two main categories by lenders: "genuine" (and therefore "deserving") or "non-genuine" ("undeserving"). Non-genuine borrowers were often discussed as having addiction problems (e.g., gambling, drugs) and being serial borrowers (having multiple loans and defaulting on repayments regularly). In contrast, self-discipline was seen as a key positive attribute of a genuine borrower: "When someone walks through the door, and they come in organized, you know that they are going to be probably a good prospect to pay you the money back. The ones who are not are genuine are the ones you don't want to deal with ... And this flies in the face of what the financial counsellors, the consumer law advocates, and all these people who say, you're just trying to get money off of them" (payday lender, Brisbane, Australia).

The last part of this utterance about "just trying to get money off of them" is a reference to the criticism that payday lenders charge excessive interest rates and fees, which is seen as legitimate by the lenders, owing to the higher risk of loan default for this group. What is interesting, at least from the perspective of borrowers, is that many of them do not consider the practice immoral. In this way, the higher cost is part of the moral norm for both borrowers and lenders in this particular market. But it is a norm borne of necessity, where few options are available in the wider mixed economy of credit. From a social justice point of view it is difficult to justify "the poverty premium" that low-income borrowers pay for accessing credit. This begrudging acceptance of the high cost within the moral economy of fringe lending may be close to what Gilbert (2013) calls "disaffected consent," where consent for the high cost of the loan is being given, but it is far from an enthusiastic acceptance of the loan conditions.

The Perspective of Borrowers

The above discussion highlights the fact that borrowers are not as naive as they are portrayed by consumer advocates concerned about exploitation and predatory lending. Self-practices around poverty management include a high degree of emotional labour and a pragmatic rationality about the costs and benefits of accessing credit from different courses. Both dimensions are elaborated here.

Overall, the study found that the emotional work involved in gleaning day-to-day credit invariably took different forms, often eliciting such strong feelings that low-income householders shied away from potential financial sources. Most people felt that borrowing from family was "wrong," "odd," a "drama," a "big problem" or "causes conflict" and often took too high an emotional toll. As one borrower explained, even if a parent or sibling had some disposable income, asking for money "was extreme because ... even though I'm forty-one, you still get a lecture." Most borrowers involved in the study readily acknowledged that the financial costs of the payday loans were too high, but they also accepted it as the price if they wanted to avoid stigma, shame, or grief.

Once a person had committed to the process of obtaining a small loan, they downplayed the high financial cost. They understood they were being charged a high rate, but they were not interested in focusing on the fine details of the contract. This attitude demonstrates the limits of policy approaches such as the various empowering technologies, including "financial literacy" educational programs for low-income citizens. These programs presuppose that the subject is a rational economic actor located in a field of economic rewards and incentives (Walters, 2000: 142). As Dean (2010: 223) argues: "Among the preferred models of the neo-liberal prudential subject is the rational choice actor who calculates the benefits and costs of acting in a certain way and then acts." One problem with using this model to promote personal and economic security is that, at least in the case of poverty survival, there are considerations beyond an economic calculus and there is little time for reflexivity and information gathering. The immediacy of material deprivation is a powerful force in directing action, as is weighing up the emotional transaction costs of different options.

It would appear that – at least psychologically and emotionally – the borrower has entered into the loan contract well before appearing in the lender's doorway. The short-term financial need and minimal emotional cost overshadows the longer-term concerns about the economic risk of not being able to repay the loan: "I knew what I was doing … But at the time, my need for what I wanted was greater than my need to pay them back. I thought, I'll just worry about it when the time comes" (borrower, Brisbane, Australia).

As flagged at the beginning of the chapter, a great deal of public policy concern has centred on the predatory practices of lenders and the vulnerability of naive borrowers. Although the exact details of the loan were not clear for borrowers, they certainly understood that a payday loan was an expensive option: "You've got to recognize that you're dealing with a loan company that charges 40 or 50 per cent interest rate, but that's the risk you've got to take or that's the interest you've got to take when you want to borrow the money on a short-term and without a credit rating" (borrower, Melbourne, Australia). Hence, a dominant position adopted by borrowers was one expressed by a "pragmatic borrower": "Well I don't know … they're pretty rude – the rates they charge – but … like I said, they're there and people need them, you know, very much" (borrower, Melbourne, Australia).

This pragmatism was echoed in borrowers' understanding of the business model of lenders: "It's a bit of a rip off; in one sense I think they're exploiting people because of their unfortunate situations but they're a business, they're there to make money, aren't they?" (borrower, Brisbane, Australia). Borrowers did not approve of the rates charged, but they accepted that if they wanted small amounts of money quickly, then this was the option available and these were the conditions under which they could access it. They understood the issue of access to credit within a system's perspective and saw small lenders as being a necessary supplement to the formal economy:

> Look, to be quite honest, I think [lenders] fulfil a real need because it comes down to credit rating, it really does. For me, for most people, most people not everyone, most people are aware that their fees are much more exorbitant than any bank or normal credit card. But the second you are in the system, and that credit rating has been affected by something – I can't even get a mobile phone contract – I think the regulation is important so that, of course, people don't get taken advantage of, but I honestly think they provide a service that's needed. (Borrower, Melbourne, Australia; Banks et al., 2012)

One particularly fascinating comment came from a borrower who revealed that not only did they think the rate charged was reasonable, but that the profit margin of the lender was a reasonable explanation for the interest charged: "As far as I'm concerned if I borrow $50 and I pay back $60 or $62, $12 is not going to kill me and it helps these people out too" (borrower, Brisbane, Australia).

The wider sociocultural context in which neoliberal political rationalities are dominant and individuals are encouraged to see everything as having a market price helps explain the acceptance by this borrower of paying a premium for short-term credit. The borrower also accepts the logic of risk applied in these transactions, which is that this group of borrowers are "riskier" because they are on a low income, and therefore there is a higher chance they will default on the loan. Other borrowers went further, promoting the idea of personal responsibility more broadly: "It comes down to the individual and the choices they made. You could equally live in a place like Lismore and go to the pub or the TABs or something every day, spend $80 a day or whatever on beer and pot or whatever, that kind of thing. Or you can be making different choices" (borrower, Brisbane, Australia). This excerpt illustrates an embrace of a neoliberal subjectivity, namely, accepting responsibility for one's choices alongside an acknowledgment of a market logic.

A neoliberal discourse of freedom to choose and responsibility for personal choices were supported by many lenders in the study: "Nobody is breaking their arms and forcing them to take these loans, any more than they're forcing them to buy a particular kind of car, or you know, and when you're talking about responsibility in lending, I think we've got to come back to responsibility in borrowing. You cannot make the lender responsible for the borrower's choices" (borrower, Brisbane, Australia; Marston & Shevellar, 2010b). The idea of freedom through responsible market choices comes through strongly in this excerpt. However, the notion of choice is hotly contested in the poverty studies literature. Recent US research found that most customers perceived that they had few, if any, options other than payday loans (Elliehausen, 2009). And despite their positive attitude towards lenders, borrowers remain concerned about what they perceive to be the high cost of financial loans (Cypress Research Group, 2004). In the context of facing a choice about having the electricity cut off at home or paying a high price for a loan to cover the electricity bill, it becomes a little easier to see why people will access a high-cost loan, regardless of high fees. The relationship between lender and borrower is also an important part of any explanation of why someone decides to take out a loan. Issues of trust and respect are critically important in such relations.

These issues are also evident in the common business practice of "working out a budget" with a borrower, where there is a mixture of confession and pastoral power at play (Foucault, 1982). As a governing technology, the calculation of the household budget becomes a key factor for both borrowers and lenders in establishing trust and managing norms and expectations. It is standard practice for lenders to sit down with borrowers and go through a list of income and expenses. The business of "working out a budget" becomes a way of confessing one's debt sins to the lender and revealing what Margaret Atwood (2008) calls a "debt plot," where everyone has their own narrative about how they got into debt and how they might get out. For the lender the budget becomes a way of exercising what Foucault refers to as a form of "pastoral power." The lender sacrifices their time for what they justify as the collective good of giving borrowers the capacity to be more responsible managers of their financial selves (which of course also helps ensure the lender gets paid). From another point of view the time lenders spend with borrowers working out the initial budget of how much they can afford to borrow could be seen as the first step in entrapment, in locking in a loyal customer.

One lender involved in the study admitted that the time it takes to work out a budget with borrowers is time well invested in future profit: "Now there is a good – sure it costs money – but there is a good economical reason, financial reason to do this. Most of our loans do not make a profit on the first loan, and in fact that's throughout retail. That's a retail saying. You do not make a profit on the first sale, because you put so much into establishing your customer on the first sale, and that's the way I look at this" (lender, Gold Coast, Australia).

There is a perceived reciprocity here that affirms the moral virtue of the lender and the needs of the borrower. The borrower-lender relationship in these localized operations revolves around notions of salvation, sin, obedience, and truth. The positive relationship between lender and borrower becomes more difficult to sustain when the borrower has difficulty repaying the loan. If the loan is secured against an asset, then the lender can also exercise coercive power through threatening to seize the fridge, the car, or the television. In these situations, borrowers seek to use their relationship with the lender to buy some time before this happens (Marston & Shevellar, 2010a). They seek to renegotiate the contract with the lender when they find it difficult to repay under the original terms of the loan: "So it's again just telling people that, 'This is where I am. I'm having a hard time. I can give you just a little out of each pay cheque. It will take me a little while but I'll catch up'" (borrower, Melbourne, Australia; ibid.). Other borrowers will use advocates, such as financial counsellors, to act on their behalf to renegotiate the loan. The many relationships that can be involved in the repayment of debt reminds us of Lipsky's (1980) observation that poorer citizens can start to accrue small armies of street-level actors intervening on their behalf.

Overall, what emerges in these accounts is a networked governing dynamic with borrowers negotiating a space between the public welfare sector, the informal sector, and the private market sector in poverty governance. What these accounts also show is that, in a relatively short period of time, payday lenders have become institutionalized in the lives of people struggling with poverty on the urban fringe. This situation would suggest that it is naive to simply wish payday lenders away without considering what other strategies are required to reduce household indebtedness, while acknowledging that the dominance of neoliberal political rationalities makes it difficult for governments to assert a role beyond regulating conduct of lenders through law reform. Thus, policy actors are left concerned with this problem but trapped within an entirely inadequate framing of the policy problem: between light and heavy forms of regulation and legal remedy. What a governmentality analysis can reveal is other possibilities for policy action, as it problematizes the conception of the naive borrower and the simple binary between the state and the market. Without a richer understanding of competing rationalities, policy makers will continue to design interventions and policy solutions that seek to engage a presupposed rational economic self, one that is detached from a complex mixed economy of credit where emotions and issues of trust and respect inform everyday financial decisions. If these dimensions are not considered, policy and practice interventions, such as financial literacy, are likely to be misguided and ineffective.

Conclusion

The increasing reliance on payday lenders by low-income households reveals important insights into the governance of poverty in late modern societies. Poverty researchers and policy makers need to take fuller account of the complex causes and consequences of the increasing demand for payday lending. While their simple solutions, such as banning payday lending, may be enticing, they do not address the larger question of how to reduce the demand that results from underlying economic insecurity resulting from precarious work and welfare.

A political economy narrative of economic insecurity resulting from global capitalism has a place in identifying the contours of contemporary poverty governance, but it misses the opportunity to identify spatial differences. Payday lending and borrowing is an integral feature of the globalization of consumer credit, but it also retains a spatial embeddedness, owing to differences in national, local, and individual financial practices. Powerful economic upheavals are reshaping the local territory, and while some are continuous with global capitalism, there are also discontinuities and contingencies that can be grasped with greater precision though use of an analytics of governmentality. Such an approach aims to elucidate the specific practices through which the conduct of borrowers and lenders is governed and how social norms about money and credit change over time. When used with ethnography (broadly understood), governmentality concepts enable the study of actors' intentions in relation to the "conduct of conduct." It is only then that we can explain how rationalities play out in practice. As Bruno Latour (cited by Walters, 2000: 150) advises, we should "follow the actors" and take seriously the way they define one another intersubjectively – or not, as the case may be. The actors in this field of practice, the fringe lenders and low-income borrowers, embrace complex and contradictory discourses and rationalities that cannot be understood without reference to developments in the mainstream financial sector, which have tended to render invisible the demand for credit among low-income groups.

ACKNOWLEDGMENTS

The study was funded by the Australian Research Council Linkage Scheme in collaboration with the National Australia Bank and Good Shepherd Youth and Family Services Victoria. I would also like to thank the editors for their insightful and helpful comments on earlier drafts of the chapter. A small amount of material from this chapter was previously published in Marston & Shevellar (2010a, 2010b) and Banks, Marston, Russell, & Karger (2012).

REFERENCES

Atwood, M. (2008). *Payback: Debt and the shadow side of wealth*. Toronto: Anansi.

Banks, M., Marston, G., Russell, R., & Karger, H. (2012). *Caught short: Final report on the experience of payday lending in Australia*. Melbourne: RMIT University.

Brady, M. (2014). Ethnographies of neoliberal governmentalities: From the neoliberal apparatus to neoliberalism and governmental assemblages. *Foucault Studies, 18*, 5–10.

Castles, F. (2004). *Future of the welfare state: Crisis myths and crisis realities*. Oxford: Oxford University Press. http://dx.doi.org/10.1093/0199270171.001.0001

Chant Link and Associates (2004) A report on financial exclusion in Australia. Available at http://www.ncrc.org/global/australAsia/Australia/AustraliaArticle6.pdf. Accessed 12 March 2013.

Clarke, J. (2006). Consumerism and the remaking of state-citizen relations in the UK. In G. Marston & C. McDonald (Eds), *Governmentality and social policy analysis* (89–106). Cheltenham, UK: Edward Elgar.

Consumer Action Law Centre (2013). What warning? Observations about mandated warnings on pay-day lender websites, Available at http://consumeraction.org.au/wp-content/uploads/2013/10/What-warning-August-2013.pdf. Accessed 24 April 2015.

Cypress Research Group (2004). *Pay-day advance: Customer satisfaction survey.* Shaker Heights, OH: Cypress Research Group.

Dean, H., & Melrose, M. (1996). Unravelling citizenship: The significance of social security benefit fraud. *Critical Social Policy, 16*(48), 3–31. http://dx.doi.org/10.1177/026101839601604801

Dean, M. (2010). *Power and rule in modern society* (2nd ed.). London: Sage.

De Parle, J. (2004). *American dream: Three women, ten kids, and a nation's drive to end welfare.* London: Penguin Books.

Elliehausen, G. (2009). *An analysis of consumers' use of pay-day loans.* Washington, DC: George Washington University School of Business.

Esping-Andersen, G. (1990). *The three worlds of welfare capitalism* (1st ed.). Cambridge: Polity Press.

Fairbanks, R. (2009). *How it works: Recovering citizens in post-welfare Philadelphia.* Chicago: University of Chicago Press. http://dx.doi.org/10.7208/chicago/9780226234113.001.0001

Foucault, M. (1982). The subject and power. *Critical Inquiry, 8*(4), 777–95. http://dx.doi.org/10.1086/448181

Foucault, M. (1991). *Discipline and punish: The birth of a prison.* London: Penguin Books.

Gilbert, J. (2013). What kind of thing is neoliberalism? *New Formations: A Journal of Culture/Theory/Politics, 80/81,* 7–22. http://dx.doi.org/10.3898/nEWF.80/81.IntroductIon.2013

Gillam, Z. (2010). *Pay-day loans: Helping hand or quicksand?* Melbourne: Consumer Action Law Centre.

Gottschalk, M. (2000). *The shadow welfare state: Labor, business, and the politics of health care in the United States.* New York: Cornell University Press.

Hacker, J. (2006). *The great risk shift: The assault on American jobs, families, health care and Retirement and how you can fight back.* Oxford: Oxford University Press.

Harris, J., & McDonald, C. (2000). Post-Fordism, the welfare state and the personal social services: A comparison of Australia and Britain. *British Journal of Social Work, 30*(1), 51–70. http://dx.doi.org/10.1093/bjsw/30.1.51

Howell, N. (2005). *High cost loans: A case for setting maximum rates?* Brisbane: Centre for Credit and Consumer Law.

Howell, N., Wilson, T., & Davidson, T. (2008). Interest rate caps: Protection or paternalism, Available at http://eprints.qut.edu.au/45483/1/CCCL-Interest-rate-caps-report-final.pdf. Accessed 24 April 2015.

Hunt, A. (2004). Getting Marx and Foucault into bed together! *Journal of Law and Society, 31*(4), 592–609. http://dx.doi.org/10.1111/j.1467-6478.2004.00305.x

Jessop, B. (1999). The changing governance of welfare: Recent trends in its primary functions, scale, and modes of coordination. *Social Policy and Administration, 33*(4), 348–59. http://dx.doi.org/10.1111/1467-9515.00157

Jordan, B., & Travers, A. (1998). The informal economy: A case study in unrestrained competition. *Social Policy and Administration*, 32(3), 292–306. http://dx.doi.org/10.1111/1467-9515.00104

Karger, H. (2005). *Shortchanged: Life and debt in the fringe economy*. San Francisco: Berrett-Koehler.

King, U., & Parrish, L. (2007). *Springing the debt trap: Rate caps are the only proven pay-day lending reform*. Durham, NC: Centre for Responsible Lending.

Lavalette, M., & Mooney, G. (1999). New Labour, new moralism: The welfare politics and ideology of New Labour under Blair. *International Socialism Journal.*, 85, 27–47.

Lavalette, M., & Pratt, A. (2006). *Social policy: Theories, concepts and issues* (3rd ed.). London: Sage. http://dx.doi.org/10.4135/9781446280386

Leyshon, A., & Thrift, N. (1997). *Money/space: Geographies of monetary transformation*. London: Routledge.

Lipsky, M. (1980). *Street-level bureaucracy: Dilemmas of the individual in public services*. New York: Russell Sage Foundation.

Lister, R. (2004). *Poverty*. Cambridge, MA: Polity.

Marston, G., and Shevellar, L. (2010a) *The experience of using fringe lenders in Queensland: A pilot study*. University of Queensland-School of Social Work & Human Services – Social Policy Unit.

Marston, G. and Shevellar, L. (2010b) *Using fringe lenders in Queensland: Report of a pilot study*, Brisbane: University of Queensland.

Marston, G., & McDonald, C. (Eds). (2009). *Analysing social policy: A governmental approach*. Cheltenham, UK: Edward Elgar.

Martin, R. (2002). *Financialization of daily life*. Philadelphia: Temple University Press.

Murphy, J., Murray, S., Chalmers, J., Martin, S., & Marston, G. (2011) *Life on welfare in Australia*. Sydney: Allen & Unwin.

NFSF (2010). Code of ethics/customer. Available At http://www.nfsf.org.au/code-of-ethics.htm. Accessed 12 March 2012.

Ong, A. (2006). *Neoliberalism as exception: Mutations in citizenship and sovereignty*. Durham, NC: Duke University Press. http://dx.doi.org/10.1215/9780822387879

Peck, J. (2010). *Constructions of neoliberal reason*. Oxford: Oxford University Press.

Pierson, P. (Ed.) (2001). *The new politics of the welfare state*. Oxford: Oxford University Press. http://dx.doi.org/10.1093/0198297564.001.0001

Piven, F., & Cloward, R. (1971). *Regulating the poor: The functions of public welfare*. New York: Pantheon Books.

Rivlin, G. (2011). *Broke USA: From pawnshops to poverty inc – how the working poor became big business*. New York: HarperCollins.

Rose, N. (1999). *Powers of freedom: Reframing political thought*. New York, Cambridge : Cambridge University Press. http://dx.doi.org/10.1017/CBO9780511488856

Schram, S. (2006). *Welfare discipline: Discourse, governance and globalization*. Philadelphia: Temple University Press.

Schram, S., Soss, J., Houser, L., & Fording, R. (2010). The third level of US welfare reform: Governmentality under neoliberal paternalism. *Citizenship Studies, 14*(6), 739–54. http://dx.doi.org/10.1080/13621025.2010.522363

Scutella, R., and Sheehan, G. (2006) *To their credit: Evaluating an experiment with personal loans for people on low income*. Fitzroy, Victoria, AUS: Brotherhood of St Lawrence.

Soss, J., Fording, R., & Schram, S. (2011). *Disciplining the poor: Neoliberal paternalism and the persistent power of race*. Chicago, London: University of Chicago Press.

Spicker, P. (2007). *The idea of poverty*. Bristol, UK: Policy Press.

Springer, S. (2012). Neoliberalism as discourse: between Foucauldian political economy and Marxian poststructuralism. *Critical Discourse Studies, 9*(2), 133–47. http://dx.doi.org/10.1080/17405904.2012.656375

Standing, G. (Ed.) (2011). *The precariat: The new dangerous class*. London: Bloomsbury Academic. http://dx.doi.org/10.7208/chicago/9780226768786.001.0001

Stegman, M. (2007). Pay-day lending. *Journal of Economic Perspectives, 21*(1), 169–90. http://dx.doi.org/10.1257/jep.21.1.169

Streeck, W., & Thelen, K. (2005). *Beyond continuity: Institutional change in advanced economies*. Oxford: Oxford University Press.

Wacquant, L. (2009). *Punishing the poor: The neoliberal government of social insecurity*. Durham, NC: Duke University Press. http://dx.doi.org/10.1215/9780822392255

Walters, W. (2000). *Unemployment and government: Genealogies of the social*. Cambridge: Cambridge University Press. http://dx.doi.org/10.1017/CBO9780511557798

Walters, W. (2006). "The end of the passing past": Towards polytemporal policy studies. In G. Marston & C. McDonald (Eds), *Governmentality and social policy analysis* (177–94). Cheltenham: Edward Elgar.

Whiteford, P. (2011). How fair is Australia's welfare state? Inside story. Available at http://inside.org.au/how-fair-is-australia%e2%80%99s-welfare-state/. Accessed 14 June 2012.

Wilson, T. (2004). The inadequacy of the current regulatory response to pay-day lending. *Australian Business Law Review., 32*(1), 193–206.

Closings

12 Governmentalities, the Ethnographic Imaginary, and Beyond

RANDY K. LIPPERT AND MICHELLE BRADY

Three key ideas have motivated this volume: neoliberalism, governmentality, and the ethnographic imaginary. The opening chapter noted that the aim was to bring together diverse works that engage with these three ideas in various ways. This objective reflected a need to move forward debates about neoliberalism, and governmentality in particular. What, then, have we learned that might inform future scholarship?

First, we can ask: what do these chapters reveal about neoliberalism? Despite covering a remarkable range of practices, including those that are geographically distant or usually invisible (e.g., at the nanolevel, in residential private spheres, or in state statistics bureaus), all chapters address this concept. As the introductory chapter outlined, a recurring criticism of some studies of governmentality is their use of neoliberalism as a master category to explain a vast array of distinct governmental programs (Rose, O'Malley, & Valverde, 2006). Others have made the contrary criticism that studies of governmentality adopt a messy view of neoliberalism that emphasizes contingency, multiplicity, combinations, and complexity. Wacquant (2012: 70) has heavily criticized these studies for suggesting that neoliberalism is "everywhere and nowhere at the same time." Furthermore, he suggests that these studies locate the process of "neoliberalization" in "the migration of 'malleable' technologies of conduct that are constantly 'realigned' and 'mutating' as they travel" (ibid.) Contrary to Wacquant's charges, the authors in this volume do not seek to analyse the "travels" of neoliberal "technologies of conduct." Instead, several chapters actually challenge the idea that we can label particular technologies of government as inherently neoliberal, thus disturbing the notion that there is a broader phenomenon of neoliberalism with known parameters and features. The chapters by

Howard, Lippert, and Valverde, in particular, reveal how neoliberal rationalities have articulated with technologies that for the most part predate neoliberalism's emergence (see also Lippert, 2010; Valverde, 1996).

Future analysis that builds on these understandings will be more challenging because it will need to avoid the common (and now somewhat tedious) aim of revealing how a largely monolithic neoliberalism encounters its alleged predecessors (e.g., social liberalism) or how inherently neoliberal technologies are used to put into practice neoliberal rationalities. Future research that seeks to understand governing practices in the domains explored in this collection and beyond must investigate these assemblages of liberal and non-liberal rationalities and the relative independence of technologies from any particular rationality. This approach is vital, too, if one is to begin to conceive of how domains, from education to urban housing to treaty negotiations to welfare provision to nanotechnology development, could be governed differently and in more progressive and inclusive ways. As Mitchell (2006: 404) has argued, the alternative approach of portraying contemporary governance as "seamless assemblages of neoliberal power" makes it hard to imagine how things could be otherwise. When taken together, the chapters in this volume, with their close attention to concrete governing practices, show why this alternative is now unacceptable, much less a viable path forward for scholars.

Second, the contributors highlight the extent to which an ethnographic imaginary is productive when combined with an analytics of governmentality. This imaginary has been remarked upon in diverse ways in this collection. As noted in chapter one, such an imaginary entails traditional understandings of ethnography in the sense of long-term, intensive fieldwork, but also a more broadly understood sensitivity to context. Diversity is evident not only in the contributors' data collection methods and sources, but also in the kinds of contributions the authors made. For example, Li brings to the collection a crucial distinction between ethnography *in* and ethnography *of* government. Her chapter underscores that ethnography has sometimes been deployed *within* governmental programs, a notion that is easily overlooked but vital. Her work encourages future scholars to be reflexive, namely, to recognize that, like the subjects of government they analyse, they too are located within governmental assemblages of one kind or another. In other words, the role of the ethnographic imaginary in governmental assemblages can sometimes be the object of study.

Most other contributors, however, engaged in ethnography of government. What this meant for the data sources and methods they used varied. While some contributors undertook extended fieldwork, others engaged with data and sources less commonly associated with ethnography. Thus, while Dorow undertook long-term participant observation within specific field sites, Akinwumi and Blomley engaged with various documents, including newspaper articles, press releases, and transcripts of consultation hearings. Other authors, including Shields, Lippert, Howard, and Marston, relied primarily on data generated through interviews or a combination of interviews and focus groups. Although there are important divergences among these scholars' aims, partially but not exclusively owing to disciplinary preferences, we nevertheless suggest they share a desire to focus their gaze on concrete manifestations of government – *governing practices* – within a specific milieu.

The foregoing chapters serve as models for approaching an understanding of concrete practices of governing within particular domains, but they do not establish a new methodological approach. Rather, they show that an ethnographic imaginary is best thought of as a sensitivity to concrete practices in context, not necessarily as a method or a methodology. This sensitivity contrasts with some previous studies of governmentality that are perhaps less concerned with understanding the situated context within which problematizations or forms of subjectivity develop or with how strategies for governance become assembled. Authors in this collection demonstrate concern with situated actors' reflections and accounts as either subjects of governance or as those who govern the conduct of others. We suggest that future studies seeking to examine specific policies can benefit from embracing the examples of an ethnographic imaginary presented in this collection. Moreover, in showing the embrace of an ethnographic imaginary across multiple disciplines, these chapters collectively underscore its transdisciplinarity, which befits not only the character of the governmentality literature, but also the broader trend of toppling artificial disciplinary boundaries in contemporary critical scholarship (see Lippert & Walby, 2013).

Why focus on concrete manifestations of government or what we term "governing practices"? For Akinwumi and Blomley, the aim is to reveal the vector of a distinctive strategy of government that involved mobilizing the agency of the public to achieve certainty about First Nations' land claims in British Columbia. In so doing they foreground a relatively neglected issue within studies of governmentality, namely, partisan disputes (Hindess, 1997: 261). Mitchell and Lizotte likewise respond to a recurring criticism of studies of governmentality: the relative lack of attention to failure or resistance. They adopt an ethnographic imaginary to focus on those individuals who are the target of a particular program of governance. Focusing on the role of philanthropic organizations in education reform in the United States, they seek to understand interactions between these organizations and community partners. Thus, they move beyond the ways that current educational arrangements have been problematized by philanthropic organizations to the messy process of incentivizing subjects and inculcating ideas of choice, entrepreneurialism, and other elements of neoliberal rationality. For Dorow, Howard, Larner and Moreton, Lippert, and Marston, the key aim is to reveal rationalities or forms of subjectification that would otherwise remain invisible. With the exception of Howard this involved actualizing a key aim of studies of governmentality: to analyse governance beyond the state by focusing on a diverse range of organizations that govern conduct, including condo corporations, oil companies, community organizations, and payday lenders. In turn, they reveal non-liberal and "beyond neoliberal" rationalities or forms of subjectification that have received little attention as well as how rationalities coexist or intertwine in forming governmental assemblages. Future research must closely attend to these governing practices if we are to more fully understand how governing in and beyond the state operates, that is, through what agents, technologies, and forms of knowledge.

Finally, how did the chapters challenge or extend our understanding of governmentality? As noted earlier, they remind us that the governmentality analytic continues to be relevant to a wide array of disciplines. Consistent with this answer, and, as suggested in chapter one, the contributors did not gloss over differences in these understandings or present a unitary way of extending or challenging an analytics of governmentality. A minority of authors argued that we can extend governmentality by embracing the normative and structural conventions of political economy. Thus, Marston, as well as Mitchell and Lizotte, argued that there still is reason for these offspring of Foucault and Marx, respectively, to prosper in the same house if not to "get into bed together" (Hunt, 2004).

The volume also moved beyond the well-rehearsed analysis of generic responsibilized or entrepreneurialized subjects of economic government, through chapters that revealed self-sustaining, collectivized, and complexly differentiated market and non-market subjects and through others that disturbed complacent assumptions about neoliberalism's features. Some of these subjectivities were informed by rationalities with non-western origins, while others were shaped by colonial rationalities that have articulated with neoliberalism or nascent and unnamed rationalities around nanoscience.

However, the primary contribution of this collection is to show how situated actors in the present reflect on, account for, and represent existing practices of government of which they are a part; how forms of reasoning associated with neoliberalism are tethered to other rationalities; and the specific governmental assemblages that have emerged from these complex mixtures. The foregoing chapters also show that governmentality does not equate to neoliberalism – these are not interchangeable terms – as some critiques of the governmentality literature have mistakenly asserted (see Lippert, 2005).

The chapters also demonstrate the benefits of using governmentality as a "soft ... conceptual lever" (Osborne, cited in Bröckling, Krasmann, & Lemke, 2011: 16),[1] which helps us move beyond an overwhelming focus on how rationalities produce "conditions of possibility" (Collier, 2009: 96). These studies reveal instead the benefit of devoting careful attention to governing practices to uncover their myriad and mutating forms, of carefully identifying the specific logics that inform such practices, some of which might be identified as neoliberal while others clearly are not. The ethnographic imaginary should, we suggest, inform that complex endeavour going forward. Ultimately, this collection's success will be measured by the extent to which it renders it more difficult to conceive of an analysis of governmental arrangements in any current domain without embracing an ethnographic imaginary.

NOTE

1 The original is in German, which is why we rely on the quotation in Bröckling et al. (2011).

REFERENCES

Bröckling, U., Krasmann, S., & Lemke, T. (Eds). (2011). *Governmentality: Current issues and future challenges*. New York: Routledge.

Collier, S.J. (2009). Topologies of power: Foucault's analysis of political government beyond "governmentality." *Theory, Culture & Society*, 26(6), 78–108. http://dx.doi.org/10.1177/0263276409347694

Hindess, B. (1997). Politics and governmentality. *Economy and Society*, 26(2), 257–72. http://dx.doi.org/10.1080/03085149700000014

Hunt, A. (2004). Getting Marx and Foucault into bed together! *Journal of Law and Society*, 31(4), 592–609. http://dx.doi.org/10.1111/j.1467-6478.2004.00305.x

Lippert, R. (2005). *Sanctuary, sovereignty, sacrifice: Canadian sanctuary incidents, power, and law*. Vancouver: UBC Press.

Lippert, R. (2010). Mundane and mutant devices of power: Business improvement districts and sanctuaries. *European Journal of Cultural Studies*, 13(4), 477–94. http://dx.doi.org/10.1177/1367549410377155

Lippert, R., & Walby, K. (Eds). (2013). *Policing cities: Urban securitization and regulation in a 21st century world*. New York: Routledge.

Mitchell, K. (2006). Neoliberal governmentality in the European Union: Education, training, and technologies of citizenship. *Environment and Planning. D, Society & Space*, 24(3), 389–407. http://dx.doi.org/10.1068/d1804

Rose, N., O'Malley, P., & Valverde, M. (2006). Governmentality. *Annual Review of Law and Social Science*, 2(1), 83–104. http://dx.doi.org/10.1146/annurev.lawsocsci.2.081805.105900

Valverde, M. (1996). "Despotism" and ethical liberal governance. *Economy and Society*, 25(3), 357–72. http://dx.doi.org/10.1080/03085149600000019

Wacquant, L. (2012). Three steps to a historical anthropology of actually existing neoliberalism. *Social Anthropology*, 20(1), 66–79. http://dx.doi.org/10.1111/j.1469-8676.2011.00189.x